Synthesis Lectures on Mathematics & Statistics

Series Editor

Steven G. Krantz, Department of Mathematics, Washington University, Saint Louis, MO, USA

This series includes titles in applied mathematics and statistics for cross-disciplinary STEM professionals, educators, researchers, and students. The series focuses on new and traditional techniques to develop mathematical knowledge and skills, an understanding of core mathematical reasoning, and the ability to utilize data in specific applications.

Erdal Karapınar · Ravi P. Agarwal

Fixed Point Theory in Generalized Metric Spaces

 Springer

Erdal Karapınar
Department of Mathematics
Çankaya University
Ankara, Turkey

Ravi P. Agarwal
Department of Mathematics
Texas A&M University-Kingsville
Kingsville, TX, USA

ISSN 1938-1743   ISSN 1938-1751   (electronic)
Synthesis Lectures on Mathematics & Statistics
ISBN 978-3-031-14971-9   ISBN 978-3-031-14969-6   (eBook)
https://doi.org/10.1007/978-3-031-14969-6

This Springer imprint is published by the registered company Springer Nature Switzerland AG
The registered company address is: Gewerbestrasse 11, 6330 Cham, Switzerland

Erdal Karapınar: To my wife Senem Pinar and our children Can and Ulaş Ege.

Ravi P. Agarwal: To my wife Sadhna.

Preface

The fixed point theory is one of the major research areas in nonlinear functional analysis, as well as topology and applied mathematics. Indeed, a fixed point problem is very simple to state, and hence, a vast of real-world problems can be expressed in the framework of the fixed point. Naturally, the solution of a fixed point problem turns into the solution of a real-world problem. In other words, fixed point theory has wide application potential in almost all quantitative sciences due to its nature. Consequently, both the theoretical advances and possible applications have been studied densely.

In this book, we collect the basic metric fixed point results in the setting of metric spaces, as well as, in the context of the generalized metric spaces, such as b-metric spaces and partial metric spaces. The book consists of two parts: In the first part, we consider the fundamental properties of the standard metric space together with the basic fixed point theorems in the context of metric spaces. In the second part, we consider some generalization of the notion of the metric and collect some crucial fixed point theorems in this new structures. In particular, we focus on the fixed point theorems in the context of "partial metric spaces" and "b-metric spaces".

The book contains six chapters. The first three chapters present some preliminaries and historical notes on metric spaces and on mappings. In the second chapter, we recollect the basic notions such as metric, norm, and auxiliary functions. The third chapter is devoted to collect significant metric fixed point theorems. Chapter 4 aims to collect remarkable generalization of the metric space. Fixed point theorems in the context of b-metric spaces and partial metric spaces are discussed in Chaps. 5–6, respectively. In general, we aim to bring the historically important fixed point theorems in the metric spaces with giving the proofs of a few of them. Later, we indicate how these results can be generalized in the context of new abstract spaces, such as b-metric spaces and partial metric spaces. We avoid the proof of all theorems and corollaries due to analogy.

Ankara, Turkey
Kingsville, USA
July 2022

Erdal Karapınar
Ravi P. Agarwal

Acknowledgements We would like to express our thanks to all researchers who have contributed significantly to Fixed Point Theory and its Applications.

Contents

Acronyms

$\mathbb{Z}$	Integers
$\mathbb{N}$	Natural numbers
$\mathbb{N}_0$	Nonnegative integers
$\mathbb{R}$	Real numbers
$\mathbb{R}_0^+$	Nonnegative real numbers
X	A nonempty set
(X, d)	Metric space
$(X, \|.\|)$	Normed space
T	A self-mapping on X
T^n	nth iteration of T
$Fix(T)$	The set of all fixed point of T in X
$Fix(T^n)$	The set of all fixed point of T^n in X
ϕ	Comparison function defined on $\mathbb{R}_0^+$
ψ	c-Comparison function defined on $\mathbb{R}_0^+$
Φ	The class of comparison function
Ψ	The class of c-comparison function
Ψ_b	The class of b-comparison function

Part I
Fixed Point Theorems in the Framework of Metric Spaces

In this first part, we aim to recollect and discuss the classical fixed point theorems as well as the recently published interesting results in the setting of (standard) metric spaces.

Fixed point theory can be described as a framework for researching and investigating the existence of the solution of the equation $f(p) = p$ for a certain self-mapping f that is defined on a non-empty set X. As is expected, here, p is called the fixed point of the mapping f. On the other side, we may re-consider the fixed point equation $f(p) = p$ as $T(p) = f(p) - p = 0$ and, accordingly, finding the zeros of the mapping T and finding the fixed point of f becomes an equivalent statement. This equivalence, not only enriches the fixed point theory but also, opens the doors to a wide range of potential applications in the setting of almost all quantitative sciences. For example, let us consider one of the classical open problems of number theory, finding perfect numbers: Let p be a self-mapping on a natural number such that $p(n)$ is the sum of all divisors of n for $n > 1$. Thus, any fixed points of the function p give a perfect number. In particular, 6 is the smallest perfect numbers, and $2^{74207280} \times (2^{74207281} - 1)$, with $44, 677, 235$ digits, is the biggest known perfect number.

The natural and the basic questions of the fixed point theory of can be listed as follows:

($Q1$) For a self-mapping f on a non-empty set X, which conditions are necessary and sufficient so that the mapping f possesses a fixed point in this structure?

($Q2$) What are the necessary and sufficient properties that a corresponding function f should be equipped with these conditions to guarantee that the fixed point equation $f(p) = p$ has a solution?

The fixed point theorem is split into three main research branches:

1. Metric fixed point theorem,
2. Topological fixed point theorem,
3. Discrete fixed point theorem.

© The Author(s), under exclusive license to Springer Nature Switzerland AG 2022
E. Karapınar and R. P. Agarwal, *Fixed Point Theory in Generalized Metric Spaces*,
Synthesis Lectures on Mathematics & Statistics,
https://doi.org/10.1007/978-3-031-14969-6_1

The origin of the metric fixed point theory draws back to Liouville [207] who proposed method of successive approximations for a solution of certain differential equations in 1837. Later, Picard [228] in 1890 developed this method systematically. Indeed, both of them implicitly used a fixed point approach to solve the differential equations that they dealt with. It is also known that the roots of the metric fixed point theory reach to Cauchy [226]. Roughly speaking, Cauchy proved the existence and uniqueness of an initial value problem $\frac{dy}{dx} = f(x, y)$, $y(x_0) = y_0$ for a continuous differentiable function f, first times. Lipschitz [226] refined the proof of Cauchy by using "Liptschitz constant". Peano [226] observed that differentiability conditions on f is superfluous. On the other hand, metric fixed point theory formally appeared in the distinguish result of Banach [56] in 1922. Indeed, the renowned result of Banach [56], known also as Banach Contraction Mapping Principle, is an abstraction of the successive approximation that is mentioned above. Banach contraction mapping principle can be easily stated as "every contraction in a complete metric space has a unique fixed point". As a historical note, we write down that this version was published by Caccioppoli [88] in 1931. hyper@itemfalse

On the other side, the origin of the topological metric fixed point theory can be sent back to the outstanding result of Poincaré [229]. The variants of Poincaré [229] were proved by Bohl [69], Brouwer [83] in 1912 and Hadamard [123] in 1910. Brouwer's notable result is the most famous one, which states, "a continuous map on a closed unit ball in $\mathbb{R}^n$ possesses a fixed point." An extension of this result is Schauder's fixed point theorem [260] of 1930, which states that "a continuous map on a convex compact subspace of a Banach space possesses a fixed point."

In 1955, Tarski [271] initiated the discrete fixed point theory by reporting a lattice-theoretical fixed point theorem. Discrete fixed point theory has been used in the economy, particularly in (Nash) equilibrium theory.

In 1974, Ćirić [97] published an interesting paper in which the investigated operators have a fixed point that is not necessarily unique. It has been called nonunique fixed points. Regarding the crucial role of the fixed point theorems in solving differential equations, we recall that not every equation has a unique solution. Naturally, not every operator can have a unique fixed point. Certain differential equations have periodic solutions; hence, the corresponding operators have periodic fixed points. Due to these reasons, many authors have been interested in the investigation of the existence of nonunique fixed points, and they have published fascinating results in this direction, see e.g. [4, 97, 121, 205, 206, 223, 281].

Metric Spaces

2

The notion of the metric can be considered as a generalization of two point distance that was contrived systematically first by Euclid. In the modern mathematical set-up, Maurice René Frechét [116] is the first mathematician who axiomatically formulated the notion of metric space, under the name of L-space. On the other hand, first Felix Hausdorff [129] used the term "metric space" although he mainly focused on the role of point-sets within abstract set theory.

Throughout the book, all sets are presumed nonempty.

Definition 2.1 A distance function over a set X, namely, $d : X \times X \to [0, \infty)$, is called metric, or usual metric or standard metric if

(d_1) $d(x, y) = d(y, x) = 0 \Longrightarrow x = y$;
(d_2) $d(x, x) = 0$;
(d_3) $d(x, y) = d(y, x)$;
(d_4) $d(x, z) \leq d(x, y) + d(y, z)$;

for each $x, y, z \in X$. In particular, the pair (X, d) is called metric space or usual metric space or standard metric space.

Note that in the literature, the single letter X represents the metric space, implying the pair (X, d).

Before giving the basic metric examples that are used in the basic analysis problems, we fixed some fundamental notations as follow: The letters $\mathbb{R}$, $\mathbb{N}$, $\mathbb{R}_0^+$, and $\mathbb{N}_0$ represent the real numbers, positive integers, non-negative real numbers and non-negative integers, respectively. Otherwise stated, the pair (X, d) represents metric space and T is reserved for a non-self mapping on the metric space (X, d). Further, the pth iterate of T will be denoted by the expression T^p denotes, $p \in \mathbb{N}$.

© The Author(s), under exclusive license to Springer Nature Switzerland AG 2022
E. Karapınar and R. P. Agarwal, *Fixed Point Theory in Generalized Metric Spaces*,
Synthesis Lectures on Mathematics & Statistics,
https://doi.org/10.1007/978-3-031-14969-6_2

Example

Consider the functions $d_i : \mathbb{R} \times \mathbb{R} \to [0, \infty)$, $i = 1, \ldots, 5$ which are defined by

1. $d_0(x, y) = |x - y|$, (standard metric or Euclidean metric)
2. $d_1(x, y) = |\arctan x - \arctan y|$,
3. $d_2(x, y) = \frac{|x-y|}{1+|x-y|}$,
4. $d_4(x, y) = \max\{|x|, |y|\}$,
5. $d_5(x, y) = \begin{cases} 0 \text{ if } x = y, \\ 1 \text{ otherwise.} \end{cases}$

for each $x, y \in \mathbb{R}$. Then, each d_i $(i = 1, \ldots, 5)$ forms a metric on $\mathbb{R}$.

The following are the fundamental examples for the functional analysis constructions.

Example

Let $X = C[0, 1]$ be the set of all continuous $\mathbb{R}$-valued functions that are defined on the interval $[0, 1]$. We define some interesting metrics on X as follows:

1. $d_1(f, g) = \int_0^1 |f(x) - g(x)| dx$.
   Geometrically, it is the area between their graphs.
2. $d_2(f, g) = \sqrt{\int_0^1 (f(x) - g(x))^2 \, dx}$.
   The geometric interpretation of this metric is not straightforward but it is the most important one in practice, e.g. for "east squares approximation".
3. $d_\infty(f, g) = \max\{|f(x) - g(x)| : 0 \le x \le 1\}$.
   Notice that this metric yields the largest distance between the graphs, geometrically.

2.1 Basic Topological Properties of the Metric Spaces

In this section, we recollect, briefly, some of the basic topological notions and notations.

A sequence $\{x_n\}_{n=1}^\infty$, in a metric space (X, d), converges to $x^* \in X$ provided that $\lim_{n\to\infty} d(x_n, x^*) = 0$. It is also written as $\lim_{n\to\infty} x_n = x^*$ or $x_n \to x^*$ as $n \to \infty$.

A sequence $\{x_n\}_{n=1}^{\infty}$, in a metric space, (X, d) is fundamental or Cauchy provided that for each $\varepsilon > 0$, there exists $n_0 = n_0(\varepsilon)$, such that

$$d(x_n, x_m) \leq \varepsilon, \text{ for each } n, m \geq n_0.$$

We also write it $\lim_{n,m\to\infty} d(x_n, x_m) = 0$. A metric space (X, d) is said to be complete if, and only if, any Cauchy (fundamental) sequence in X converges to a point in X.

Lemma 2.1 *Suppose that a sequence $\{x_n\}$ is a Cauchy in a (not necessarily complete) metric space (X, d). If there is a subsequence $\{x_{n_k}\}$ converging to a point x^*, then the original sequence $\{x_n\}$ also converges to x^**

Proof We must show that for any given $\varepsilon > 0$, there is $N_0 \in \mathbb{N}$ such that $d(x_n, x^*) < \varepsilon$ for each $n \geq n_0$. Since $\{x_n\}$ is a Cauchy sequence, there is an $N_1 \in \mathbb{N}$ such that $d(x_n, x_m) < \frac{\varepsilon}{2}$ for each $n, m \geq N_1$. Since $\{x_{n_k}\}$ converges to x^*, there is a q such that $n_q \geq N := \max\{N_0, N_1\}$ and $d(x_{n_q}, x^*) \leq \frac{\varepsilon}{2}$ Accordingly, for each $n \geq N$ we find

$$d(x_n, x^*) \leq d(x_n, x_{n_q}) + d(x_{n_q}, x^*) < \frac{\varepsilon}{2} + \frac{\varepsilon}{2} = \varepsilon,$$

by the triangle inequality. □

A self-mapping T on a metric space (X, d) to itself is continuous, if for any sequence $\{x_n\}$ converges to x^*, its graph sequence (namely, $\{Tx_n\}$) converges to its image under the mapping T, that is Tx^*. In other words,

$$\lim_{n\to\infty} d(Tx_n, Tx^*) = 0 \text{ whenever } \lim_{n\to\infty} d(x_n, x^*),$$

or $\lim_{n\to\infty} Tx_n = T\left(\lim_{n\to\infty} x_n\right)$.

In what follows, we recall the very interesting concept of "orbitally continuous" which is also known as "continuous orbitally":

Definition 2.2 A mapping T on a metric space (X, d) is said to be orbitally continuous (or "continuous orbitally") if

$$\lim_{i\to\infty} T^{n_i}(x) = z \text{ implies } \lim_{i\to\infty} T(T^{n_i}(x)) = Tz.$$

We shall say that a metric space (X, d) is called T-orbitally complete (or simply, orbitally complete) if each fundamental sequence of the form $\{T^{n_i}(x)\}_{i=1}^{\infty}$, $x \in X$ converges in metric space (X, d).

Remark 2.1 If a mapping T is orbital continuous, then any iteration of it (T^m for any $m \in \mathbb{N}$), is orbital continuous.

A point x^* is called a periodic point of a function T of period m if $T^m(x^*) = x^*$, where $T^0(x) = x$ and $T^m(x)$ is iteratively defined by $T^m(x) = T(T^{m-1}(x))$. The set $\text{Fix}_X(T)$ indicate the set of all fixed point of T on X.

For a metric space (X, d), we say that the self-mapping T on X is called

(i) Lipschitz continuous if for each pairs $(x, y) \in X \times X$, there is a nonnegative real number $L \geq 0$ such that
$d(Tx, Ty) \leq Ld(x, y)$;
(ii) contraction if it is Lipschitz continuous and $0 \leq L < 1$;
(iii) nonexpansive if for each pairs $(x, y) \in X \times X$,
$d(Tx, Ty) \leq d(x, y)$;
(vi) isometry if for each pairs $(x, y) \in X \times X$,
$d(Tx, Ty) = d(x, y)$;
(v) noncontractive if for each pairs $(x, y) \in X \times X$,
$d(Tx, Ty) \geq d(x, y)$;
(vi) expansive if for each distinct $x, y \in X$,
$d(Tx, Ty) > d(x, y)$;
(vii) contractive if for each distinct $x, y \in X$,
$d(Tx, Ty) < d(x, y)$ for each distinct $x, y \in X$;

Here, L is called Lipschitz constant.

Two metrics d and δ, defined on X, are called topological equivalent if $\tau_d = \tau_\delta$. On the other hand, the metrics d and δ, defined on X, are called metric equivalent if there exist $C_1, C_2 > 0$ such that $C_1 d(x, y) \leq \delta(x, y) \leq C_2 d(x, y)$, for each pairs $(x, y) \in X \times X$.

Lemma 2.2 *For any nonempty set X, let Y be any nonempty subset of X. Suppose that (X, d) is a metric space. Let $d_Y := d|_{Y \times Y}$ be the restriction of the metric d to the subset Y. Then, the pair (Y, d_Y) forms a metric space.*

The proof is straightforward and hence we skip it.

Let X be a topological space, Y be a subset of X, and $O := \{O_i : i \in I\}$ be a collection of open sets in X. We say that O is an open cover of $Y \subset X$ whenever $Y \subset \bigcup_{i \in I} O_i$. If a subcollection $\mathcal{U} := \{O_i : i \in J\}$, for some $J \subset I$, still covers Y, then $\mathcal{U}$ is called a a subcover of Y. In addition, $\mathcal{U}$ is called a finite subcover if the index set J is finite.

Next, we give the formal definition of compact metric space.

Definition 2.3 If every open cover of a metric space (X, d) has a finite subcover, then the corresponding metric space (X, d) is called compact. In addition, we say that a metric space

(X, d) is sequentially compact if each sequence of points in X has a convergent subsequence converging to a point in X.

Note that in metric space (X, d), the compactness of X is equivalent to the following statement: Each sequence in X has a convergent subsequence.

Lemma 2.3 *Every compact metric space is complete.*

Proof Let (X, d) be a compact metric space and let a sequence $\{x_n\}$ be Cauchy. Owing to the fact that (X, d) is compact, there is a subsequence $\{x_{n_k}\}$ converging to a point x^*. By Lemma 2.1, the sequence $\{x_n\}$ also converges to the same point x^*. Hence, all Cauchy sequences converge, and X must be complete. □

Definition 2.4 (*Krasnosel'skiĭ* [200]) If a mapping T on a metric space (X, d) satisfies

$$\lim_{n \to \infty} d(T^n x, T^{n+1} x) = 0,$$

for each $x \in X$, then it is called asymptotically regular.

The following example indicates that being asymptotically regular for a sequence is not sufficient to conclude it as a Cauchy sequence:

Example

Let $X = \mathbb{R}$ with a standard Euclidean metric $d(x, y) = |x - y|$ and $\{x_n\}$ be a sequence with $x_n = \log n$. Notice that

$$d(x_n, x_{n+1}) = \log \left(1 + \frac{1}{n} \right) \to 0 \text{ as } n \to \infty.$$

On the other hand, $\{x_n\}$ is not a Cauchy sequence since $\lim_{n \to \infty} \log n \to \infty$.

2.2 Connection Between the Normed Spaces and Metric Spaces

In this section, we shall briefly discuss the connection between the normed spaces and metric spaces.

First, we recall the notion of vector space. Let X be a nonempty set. We say that X is a vector space if it is closed under scalar multiplication and finite vector addition. One of the well-known and remarkable examples of vector space is n-dimensional Euclidean space

$\mathbb{R}^n$. In this construction, each element is represented by a list of n real numbers, scalars are real numbers (in the general case, the scalars are members of a field $\mathbb{F}$), scalar multiplication is multiplication on each term separately, and addition is componentwise. We generally say that X is a vector space over $\mathbb{F}$.

In what follows, we recall the concept of normed space.

Definition 2.5 A normed space is a pair $(X, \|\cdot\|)$ of a vector space X and a function $\|\cdot\| : X \rightarrow [0, \infty)$, called the norm on X, such that

1. For all $x \in X$, we have $\|x\| = 0$ if and only if $x = 0$;
2. For all scalars λ and all $x \in X$, we have that $\|\lambda x\| = |\lambda| \|x\|$;
3. $\|x + y\| \leq \|x\| + \|y\|$ for each $x, y \in X$, (the triangle inequality).

Commonly, X itself is named as a normed space by implicitly assuming that there is a norm $\|\cdot\|$ on X. In some sources, the normed space is mentioned as normed linear system.

Notice that for a normed space X, a function $d : X \times X \rightarrow [0, \infty)$, defined as,

$$d(x, y) := \|x - y\|, \text{ for each } x, y \in X, \qquad (2.1)$$

forms a metric on X. Thus, all the notions that are defined for metric spaces can be applied to normed linear spaces. We shall also consider the following metric that is obtained by a norm:

$$d_b(x, y) := \frac{\|x - y\|}{1 + \|x - y\|}, \text{ for each } x, y \in X. \qquad (2.2)$$

Note that the values of new metric $d_b(x, y) < 1$ for each $x, y \in X$.

We underline that the metric associated with a norm (2.1) has two crucial properties:

1. Translation invariance: $d(x, y) = d(x + z, y + z)$ for each $z \in X$;
2. Homogeneity: $d(cx, cy) = |c| d(x, y)$ for each $c \in \mathbb{F}$, where $\mathbb{F}$ can be $\mathbb{R}$ or $\mathbb{C}$.

In normed space, the closed ball is defined in an analogous way as in the metric case: $K := \{x \in X : \|x\| \leq 1\}$. A subset C of normed space X is called convex if $\lambda x + (1 - \lambda)y \in C$, for each $x, y \in C$ and real number $\lambda \in [0, 1]$. In other words, the subset C is convex means the line segment joining any two points in the set lies in the set. As a first observation, the unit ball is convex.

Example

Let $X = \mathbb{R}^n$ be a vector space. Here, X forms a normed space equipped with the norm

$$\|x\| := \sqrt{\sum_{i=1}^{n} x_i^2},$$

where $x = (x_1, x_2, \ldots, x_n) \in \mathbb{R}^n$. The pair $(\mathbb{R}^n, \|\cdot\|)$ is called n-dimensional Euclidean space with Euclidean n-norm $\|.\|$. In particular, 1-norm is

$$\|x\|_1 := \sum_{i=1}^{n} |x_i|.$$

Furthermore, the maximum norm is defined as

$$\|x\|_{\max} := \max\{|x_1|, |x_2|, \ldots, |x_n|\}.$$

A complete normed vector space is Banach space. Notice that every finite dimensional normed space is complete. In other words, every finite dimensional normed space forms a Banach space.

Example

Suppose $\mathbb{F}$ is either $\mathbb{R}$ or $\mathbb{C}$. For a sequence $x : \mathbb{N} \to \mathbb{F}$, we define a norm

$$\|x\|_p := \left(\sum_{i=1}^{\infty}(x(i))^2\right)^{\frac{1}{p}} \text{ if } 1 \leq p < \infty,$$

$$\|x\|_{\infty} := \sup_{i \in \mathbb{N}} |x(i)| \text{ if } p = \infty.$$

Here, $\ell^p = \ell^p(\mathbb{F}) := \{x : \mathbb{N} \to \mathbb{F} : \|x\|_p < \infty\}$ forms a normed space which is complete for each $p \in [1, \infty]$.

Note that $\|x\|_p$ does not form a norm for $0 < p < 1$.

2.3   Some Interesting Auxiliary Functions

In this section, we recall some interesting auxiliary functions that play crucial roles in the metric fixed point theory. The usage of auxiliary functions in metric space is quite natural when considering the following observation: (Lipschitz continuity) $d(Tx, Ty) \leq Ld(x, y)$ for each $x, y \in X$; with the following one

$$d(Tx, Ty) \leq \theta(d(x, y)) \text{ for each } x, y \in X;$$

turns into general form, Lipschitz θ-continuity, where $\theta : [0, \infty) \to [0, \infty)$. It is evident that $\theta(t) = Lt$, $L \geq 0$ forms the initial example of the auxiliary functions.

Next, we recall the comparison functions which has been used in metric fixed point theory, effectively. Indeed, a comparison function is one of the most interesting and useful auxiliary function that plays a crucial role in metric fixed point theory. This notion is consider initially by Browder [84] and later by Rus [60, 252] and many others.

Definition 2.6 ([60, 252]) Let ϕ be a selfmapping on $\mathbb{R}_0^+$, that is, $\phi : [0, \infty) \to [0, \infty)$. The function ϕ is called comparison if it fulfills the followings:

(i)   it is increasing;
(ii) $\phi^n(t) \to 0$ as $n \to \infty$ for each $t \in [0, \infty)$,

where ϕ^n is the nth iterate of ϕ.

The letter Φ denotes the set of all (c)-comparison functions $\phi : [0, \infty) \to [0, \infty)$. An immediate basic example of such mappings is

$$\phi(t) = \frac{kt}{n},$$

where $k \in [0, 1)$ and $n \in \{2, 3, \ldots\}$. We refer to [6, 60, 84, 251, 252] for further examples and properties of comparison functions. In the following lemma, the crucial properties of the comparison functions will be given.

Lemma 2.4 ([60, 251, 252]) *For a comparison function* $\phi : [0, \infty) \to [0, \infty)$, *we have*

(i) ϕ *is continuous at* 0;
(ii)   *for any intercalation* ϕ^k *of* ϕ, $k \geq 1$, *forms a comparison function, too;*
(iii) $\phi(t) < t$ *for any* $t > 0$.

Next, we remember the (c)-comparison functions:

Definition 2.7 Let ψ be a selfmapping on $\mathbb{R}_0^+$, that is, $\psi : [0, \infty) \to [0, \infty)$. The function ψ is called (c)-comparison function if it attains the followings:

(Ψ_1) ψ is nondecreasing;

(Ψ_2) $\sum\limits_{n=1}^{+\infty} \psi^n(t) < \infty$ for each $t > 0$.

The letter Ψ denotes the set of all (c)-comparison functions $\psi : [0, \infty) \to [0, \infty)$.

Notice that the functions in Ψ are also known as Bianchini-Grandolfi gauge functions in the literature, see e.g.[66, 231, 232]. For details on the properties and examples of (c)-comparison functions, see e.g. [66, 231, 232, 251, 252]. Observe that if ϕ is a (c)-comparison function, it automatically forms a comparison function. Consequently, each of the observed of properties above are also sustained for (c)-comparison functions.

Definition 2.8 ([6, 193]) Let φ be a selfmapping on $\mathbb{R}_0^+$, that is, $\varphi : [0, \infty) \to [0, \infty)$. The function φ is called alternating distance if it fulfills the followings:

(i)   it is non-decreasing;
(ii) $\varphi(0) = 0$;
(iii) continuous.

A function $\varphi : X \to \mathbb{R}$ is said to be lower semicontinuous at x if for any sequence $\{x_n\} \subset X$ with $\lim\limits_{n \to \infty} x_n = x \in X \Rightarrow \varphi(x) \leq \liminf\limits_{n \to \infty} \varphi(x_n)$.

2.4   Bessage, Janos and Picard Operators

Let X be a non-empty set and T be a selfmapping on X. We reserve the notation T^n for the nth iteration of the mapping T, where $n \in \mathbb{N}$. We shall use the expression $Fix(T)$ to indicate the set of all fixed point of T in X. Similarly, the notation $Fix(T^n)$ shows the fixed point set of T^n.

An operator T is called Bessage operator if

$$Fix(T^n) = Fix(T) = \{x_*\},$$

for each $n \in \mathbb{N}$.

Note that $Fix(T) \subset Fix(T^n)$. Indeed, if $x_* \in Fix(T)$, then

$$T^n x_* = T(...T(x_*)) = x_*.$$

Hence, $x_* \in Fix(T^n)$. The revise direction is not true, in general.

An operator T is called Janos operator if

$$\bigcap_{i \in \mathbb{N}} T^i = \{x_*\}.$$

An operator T is called weakly Picard operator if the sequence, $\{T^n(x)\}_{n \in \mathbb{N}}$, converges for each $x \in X$ and the limit (which generally depend on x) is a fixed point of T.
An operator T is called Picard operator if it is weakly Picard operator with $Fix(T) = \{x_*\}$.

For more details, see e.g. [251, 252].

Metric Fixed Point Theory

3

Wonder is the beginning of wisdom.

Socrates

The aim of this chapter is to give a brief history of metric fixed point theory.

3.1 Banach Contraction Mapping Principle

In this section, we discuss the pioneer metric fixed point theorem that was given by Banach [56]. This outstanding result is known as the contraction mapping principle or the Banach contraction mapping principle. The main advantage of Banach's metric fixed point theorem is the following property: This theorem not only guarantees the existence and uniqueness of fixed points of certain self maps of metric spaces but also indicates how to construct an iterative sequence that provides the desired fixed points. It is worth mentioning that this famous fixed point theorem was formulated in his thesis in 1920 and published in 1922 in the setting of normed linear spaces (not metric spaces). As we mentioned in the previous section, Banach Contraction Mapping Principle is not the first fixed point theorem in the literature but the most interesting fixed point theorem in the context of metric fixed point theory. Indeed, Brouwer gave the first result, which only guarantees the existence of the fixed point. Unfortunately, Brouwer's fixed point theorem does not explain how to get the guaranteed fixed point and how to ensure the uniqueness of this mentioned fixed point.

Theorem 3.1 ([83]) *If K is a closed ball in $\mathbb{R}^n$ and $T : K \to K$ is a continuous map then T possesses a fixed point in K.*

© The Author(s), under exclusive license to Springer Nature Switzerland AG 2022   15
E. Karapınar and R. P. Agarwal, *Fixed Point Theory in Generalized Metric Spaces*,
Synthesis Lectures on Mathematics & Statistics,
https://doi.org/10.1007/978-3-031-14969-6_3

Notice that this theorem is classified as a fixed point theorem in topological fixed point theory. We underline the fact that Brouwer's fixed point theorem concludes only the existence of a fixed point. The weakness of this theorem can be formulated by the following questions:

(W1) Is this fixed point unique?
(W2) How to find this fixed point (what exactly is it)?

On the other hand, Banach Contraction Mapping Principle, not only guarantee both the existence and uniqueness, but also show how you get the precise solution.

For the sake of completeness, we shall recall some basic notions here, and then we state Banach's famous fixed point theorem. Note that the notion of Lipschitz continuity sustains in the setting of normed spaces, either. In particular, for a non-self-mapping T, defined on a normed space $(X, \| \cdot \|)$, is called Lipschitzian (or Lipschitz continuous) if there is a constant $L \geq 0$ so that

$$\|Tx - Ty\| \leq L\|x - y\|, \quad \text{for each } x, y \in X. \tag{3.1}$$

The smallest nonnegative real number L that fulfills (3.1) is called the Lipschitz constant. It is evident that Lipschitzian maps are continuous. Recall also that if $L < 1$, then the mapping T is called contraction.

Now, we can easily state Banach Contraction Mapping Principle: Every contraction in a complete normed space possesses a unique fixed point. Caccioppoli [88] characterized the Banach Contraction Mapping Principle in the setting of complete metric space. In the recent literature, the metric version is known as Banach Contraction Mapping Principle. Throughout the manuscript, we prefer to call "Banach Fixed Point Theorem" which is formally and more precisely stated below:

Theorem 3.2 (Banach Fixed Point Theorem) *A selfmapping T, on a complete metric space (X, d), has a unique fixed point x^*, if there is a constant k in $[0, 1)$ so that*

$$d(Tx, Ty) \leq k\, d(x, y), \tag{3.2}$$

for each pair of points $(x, y) \in X \times X$. In particular, the recursive sequence $\{T^n x_0\}$, for each $x_0 \in X$, converges to x^*. In addition, for each $n \in \mathbb{N}$, we have the following estimation:

$$d(x^*, T^n x_0) \leq \frac{k^n}{1 - k} d(x_0, T x_0), \tag{3.3}$$

for any $x_0 \in X$.

Proof Take an arbitrary $x \in X$ an let $x_0 := x$. We construct an iterative sequence for this initial value so that $x_{n+1} := T x_n$ for each $n \in \mathbb{N}_0$. On account of the contraction condition, by substituting $x = x_n$ and $y = x_{n+1}$, we find

$$d(x_{n+1}, x_{n+2}) = d(Tx_n, Tx_{n+1}) \le kd(x_n, x_{n+1}) \le k^2 d(x_{n-1}, x_n)$$
$$\cdots \tag{3.4}$$
$$\le k^n d(x_0, x_1),$$

for any $n \in \mathbb{N}$. Attendantly, we find that the sequence $\{d(x_{n+1}, x_n)\}$ is bounded from below by 0 and it is non-increasing.

Further,

$$d(x_n, x_{n+r}) \le \sum_{i=n}^{n+r-1} d(x_i, x_{i+1}) \le \sum_{i=n}^{n+r-1} k^i d(x_0, x_1) \le \frac{k^n}{1-k} d(x_0, x_1). \tag{3.5}$$

Since $k < 1$, the geometric series yields that $d(x_n, x_{n+r}) \to 0$ as $n, r \to \infty$. That is, the sequence $\{x_n\}$ is Cauchy and accordingly it converges to some $x^* \in X$. Since, T is (necessarily) continuous, we find

$$Tx^* = T(\lim_{n \to \infty} x_n) = \lim_{n \to \infty} Tx_n = \lim_{n \to \infty} x_{n+1} = x^*, \tag{3.6}$$

that is, x^* is the fixed point of T. Uniqueness can be easily obtained by the method of *Reductio ad Absurdum*. Suppose, on the contrary, that x^* and z^* be two distinct fixed points of T. Concomitantly, by (3.2) we have

$$d(x^*, z^*) = d(Tx^*, Tz^*) \le kd(x^*, z^*) < d(x^*, z^*), \tag{3.7}$$

a contradiction. As a result, x^* is the unique fixed point of T.

Letting $r \to \infty$ in the inequality (3.5), we derive that

$$d(x^*, T^n x_0) = \lim_{r \to \infty} d(x_n, x_{n+r}) \le \frac{k^n}{1-k} d(x_0, x_1) = \frac{k^n}{1-k} d(x_0, Tx_0).$$

$\square$

An alternative proof of Banach Contraction Principle: On account of fixed point equation (3.2), for each $x \in X$, we have

$$d(Tx, T^2 x) \le k\, d(x, Tx). \tag{3.8}$$

By adding the term $d(x, Tx)$ to the both sides of the inequality (3.8), we find

$$d(x, Tx) + d(Tx, T^2 x) \le d(x, Tx) + k\, d(x, Tx).$$

After a simple evaluation, we derive, from the above equation, that

$$d(x, Tx) \le (1-k)^{-1}[d(x, Tx) - d(Tx, T^2 x)] = \varphi(x) - \varphi(Tx),$$

where the function $\varphi : X \to [0, \infty)$ is defined as

$$\varphi(x) = (1 - k)^{-1} d(x, Tx), \text{ for } x \in X.$$

By using this observation, we shall prove that $\{T^n x\}$ is a fundamental (Cauchy) sequence. For this purpose, take $m, n \in \mathbb{N}$ with $n < m$, the triangle inequality yields that

$$d(T^n x, T^{m+1} x) \le \sum_{i=n}^{m} d(T^i x, T^{i+1} x) \le \varphi(T^n x) - \varphi(T^{m+1} x).$$

By setting $n = 1$ and letting $m \to \infty$ we deduce that

$$\sum_{i=1}^{\infty} d(T^i x, T^{i+1} x) \le \varphi(Tx) < \infty.$$

Accordingly, we conclude that $\{T^n x\}$ is a Cauchy sequence. The rest of the proof is verbatim of the proof above.

On the other hand, for the estimation, we consider

$$d(T^n x, T^{m+1} x) \le \varphi(T^n x) - \varphi(T^{m+1} x).$$

Taking $m \to \infty$ we find that

$$d(T^n x, x^*) \le \varphi(T^n x) = (1 - k)^{-1} d(T^n x, T^{n+1} x).$$

On account of the inequality $(1 - k)^{-1} d(T^n x, T^{n+1} x) \le \frac{k^n}{1-k} d(x, Tx)$, we conclude

$$d(T^n x, x^*) \le \frac{k^n}{1 - k} d(x, Tx).$$

This provides an estimate on the rate of convergence for the sequence $\{T^n x\}$ which depends only on $d(x, Tx)$. □

Example

Suppose $X = \mathbb{R}$ endowed with a standard Euclidean metric $d(x, y) = |x - y|$. Let $T : X \to X$ is defined by $Tx = x + c$ for each $c \in \mathbb{R}$ with $c \ne 0$. Attendantly, we derive $d(Tx, Ty) = |(x + c) - (y + c)| = |x - y|$ and $d(x, y) = |x - y|$. Thus, T forms a Lipschitzian map with $k = 1$. On the other hand, T is fixed point free and hence the contraction k can not be extended to 1.

Example

Suppose a selfmapping T on $\mathbb{R}$ is defined by $T(x) = 1 - x$, for each $x \in \mathbb{R}$. Then, T is not a contraction, has a unique fixed point. On the other hand, $T^2(x) = x$, for each $x \in \mathbb{R}$ and hence $Fix(T) = \mathbb{R}$.

Example

A selfmapping $T : [0, 2] \to [0, 2]$ is constructed by $T(x) = \begin{cases} 0 & \text{if } x \in [0, 1], \\ 1 & \text{if } x \in (1, 2]. \end{cases}$ It is clear that T is discontinuous at 1. Accordingly, it does not form a contraction. Notice that $T^2 x = 0$ for each $x \in [0, 2]$. Consequently, iteration T^2 is a contraction on $[0, 2]$ and desired fixed point is 0.

In some old sources, the renowned Banach fixed point theorem is also known as Picard-Banach. In some other sources, it is also called the Banach-Caccioppoli theorem. Indeed, the original theorem of the Banach was in the setting of normed spaces and it was revisited by Caccioppoli [88] in the setting of complete metric spaces.

The following immediate extension also belongs to Caccioppoli [88].

Theorem 3.3 (Caccioppoli [88]) *We presume that T is a selfmapping on a complete metric space (X, d). For each $n \geq 1$, if there is a constant c_n so that*

$$d\left(T^n(x), T^n(y)\right) \leq c_n d(x, y),$$

for each $x, y \in X$, where $\sum_{n=1}^{\infty} c_n < \infty$, then, T possesses a unique fixed point.

The proof can be obtained by a slight modification of Banach's proof. Indeed, by following the same notations, we have

$$d(x_{n+1}, x_{n+2}) = d(Tx_n, Tx_{n+1}) = d(T^n x_0, T^{n+1} x_0) \leq c_n d(x_0, x_1),$$

for each $n \in \mathbb{N}$. Furthermore, we can easily show that the sequence $\{x_n\}$ is Cauchy:

$$d(x_n, x_{n+r}) = \sum_{i=n}^{r} d(T^i x_0, T^{i+1} x_0) = \sum_{i=n}^{r} d(T^i x_0, T^{i+1} x_0) \leq \sum_{i=1}^{\infty} c_i < \infty.$$

We skip the rest since it is verbatim of Banach's proof.

In 1957, the following theorem was proposed by Kolmogorov and Fomin [[203], p. 501] as a generalization of Banach's fixed point.

Theorem 3.4 (Kolmogorov and Fomin [203]) *If T is a continuous mapping of a complete metric space X into itself, so that the mapping T is a contraction for some positive integer n, then the equation $Tx = x$ has one and only one solution.*

In 1965, Chu and Diaz [95] reported that the continuity assumption in Theorem 3.4 is superfluous by providing the following example:

Example

(Chu and Diaz [95]) Let X be the Banach space of all real valued continuous functions, $C([0, l])$, on the closed interval $0 \le x \le 1$, with the norm $\|f\| = \sup_{x \in [0,1]} |f(x)|$. Consider the linearly independent elements of $C([0, l])$,

$$e^x, 1, x, x^2, x^3, \ldots,$$

and extend this linearly independent set to a Hamel basis H (see e.g. Dunford and Schwartz [[109], p. 361). The transformation T is defined on H, as follows:

$$T(h) = \begin{cases} \frac{1}{2} \cdot 1, & \text{if } h = e^x, \\ \frac{1}{2} \cdot e^x, & \text{if } h = 1, \\ \frac{1}{2} \cdot h, & \text{otherwise.} \end{cases}$$

In particular, $T(x^n) = \frac{1}{2} x^n$ for each $n \in \mathbb{N}$. On account of H being a Hamel basis of $C([0, l])$, the domain of T can be extended from H to all of $C([0, l])$ as

$$T(f) = \sum_{i=1}^{n} c_i T(h_i) \text{ whenever } \sum_{i=1}^{n} c_i h_i,$$

where $c_i \neq 0$ for each $i \in \{1, \ldots, n\}$, and $h_i \in H$ for each $i \in \{1, \ldots, n\}$. Accordingly, we conclude that $T(0) = 0$ is a fixed point and $T^2(f) = \frac{1}{4} I$ is a contraction. Notice that T is not continuous at e^x. More precisely,

$$\lim_{n \to \infty} T(e^x) = \lim_{n \to \infty} T\left(1 + \sum_{i=1}^{\infty} \frac{x^n}{n!}\right) = e^x - \frac{1}{2} \neq \frac{1}{2} = T(e^x).$$

Theorem 3.5 (Chu and Diaz [95]) *Let T be a (single valued) function defined on a complete metric space (X, d). If, for some integer n, the function T^n forms a contraction, then T has a unique fixed point.*

Proof On accounts of Banach's fixed point theorem T^n has a unique fixed point x^*. Accordingly, $T^n x^* = x^*$ yields that

$$T^{n+1} x = T(T^n x) = Tx.$$

Consequently, Tx^* is also a fixed point of T^n. Since the fixed point of T^n is unique, it must be the case that $Tx^* = x^*$. Furthermore, if $Ty^* = y^*$, then we have $T^n y^* = y^*$. Again by the uniqueness we conclude that $y^* = x^*$.   □

The following is a generalization of Theorem 3.5.

Theorem 3.6 (Chu and Diaz [95]) *Let X be any nonempty set, equipped with a metric d, and T be a single valued function defined on X into itself. Suppose that, for some positive integer n, the function T^n has a unique fixed point x^*. Then T also has a unique fixed point, namely x^*.*

The proof is easily derived from the proof of Theorem 3.5. Note that in Theorem 3.6, X is just any non-empty set. If X is equipped with a metric, then Theorem 3.5 covers Theorem 3.6.

Next, we recall the definition and theorem of well-posed for fixed point.

Definition 3.1 Let (X, d) be a metric space and $Y \subset X$ a nonempty subset. The fixed point problem is well posed for an operator $T : Y \to X$ if and only if

(WP_1) $Fix(T) = \{x^*\}$;
(WP_2) If $\{y_n\}$ in Y and $d(y_n, Ty_n) \to 0$ as $n \to \infty$, then $y_n \to x^*$ as $n \to \infty$.

The condition $(WP2)$ can be expressed as follow:

(WP_2') Let $\{\varepsilon_n\}$ be a sequence of positive numbers so that $\lim_{n \to \infty} \varepsilon_n = 0$ and $\{y_n\}$ in Y. If $d(y_{n+1}, Ty_n) \leq \varepsilon_n$ as $n \in \mathbb{N}$, then $\lim_{n \to \infty} y_{n+1} = x^*$.

Theorem 3.7 *The fixed point problem is well posed for every contraction in complete metric spaces.*

Proof Due to Banach Contraction Principle, T has a unique fixed point, in particular, $T^n x \to x^*$. By letting $y_0 = x$, we find that

$$d(T^{m+1}x, y_{m+1}) \leq d(T(T^m x), T y_m) + d(T y_m, y_{m+1})$$
$$\leq kd(T^m x, y_m) + \varepsilon_m$$
$$\leq k\left[d(T(T^{m-1}x), T y_{m-1}) + d(T y_{m-1}, y_m)\right] + \varepsilon_m$$
$$\leq \sum_{i=0}^{m} k^{m-i} \varepsilon_i.$$

On the other hand, by the triangle inequality, we have

$$d(y_{m+1}, x^*) \leq d(y_{m+1}, T^{m+1} x) + d(T^{m+1} x, x^*)$$
$$\leq \sum_{i=0}^{m} k^{m-i} \varepsilon_i + d(T^{m+1} x, x^*).$$

Since $\lim_{m \to \infty} d(T^{m+1} x, x^*) = 0$, letting $m \to \infty$, we deduce that

$$\lim_{m \to \infty} y_{m+1} = x^*.$$

$\square$

Definition 3.2 We say that a self-mapping $T : X \to X$ on a metric space (X, d) is asymptotically regular *at a point* $x \in X$ if $\lim_{n \to \infty} d(T^n x, T^{n+1} x) = 0$

3.2   Further Consequences of Banach Fixed Point Theorem

We shall classify the extension of the Banach Contraction Principle in two main streams: Linear extensions and nonlinear extensions.

3.2.1   Linear Extensions of Banach Contraction Principle

The proofs of the results in the class of linear extensions of the Banach Contraction Principle mimic Banach's proof. For this reason, we shall not give any proofs in this part but only recollect several of the different contraction type mappings here.

(L1)   [Kannan [152]] There is a $\alpha \in [0, \frac{1}{2})$ so that

$$d(Tx, Ty) \leq \alpha [d(x, Tx) + d(y, Ty)], \quad \forall x, y \in X.$$

(L1a) Another variant of (L1): There is a $\alpha, \beta \in [0, 1)$ with $\alpha + \beta < 1$ so that

$$d(Tx, Ty) \leq \alpha d(x, Tx) + \beta d(y, Ty), \quad \forall x, y \in X.$$

(L2)   [Rus, Ćirić, Reich [237–239, 252]] There are α, β, γ with $0 \leq \alpha + \beta < 1$ so that

$$d(Tx, Ty) \leq \alpha d(x, y) + \beta[d(x, Tx) + d(y, Ty)], \quad \forall x, y \in X.$$

(L2a) Another variant of (L2): There are α, β, γ with $0 \leq \alpha + \beta + \gamma < 1$ so that

$$d(Tx, Ty) \leq \alpha d(x, y) + \beta d(x, Tx) + \gamma d(y, Ty), \quad \forall x, y \in X.$$

(L2b) Another variant of (L2): There are α, β, γ with $0 \leq \alpha < \frac{1}{3}$ so that

$$d(Tx, Ty) \leq \alpha \left[d(x, y) + d(x, Tx) + d(y, Ty)\right], \quad \forall x, y \in X.$$

(L3)   Bianchini [66] There exists a number $k, 0 \leq k < 1$, so that, for each $x, y \in X$,

$$d(Tx, Ty) \leq k \max\{d(x, Tx), d(y, Ty)\}$$

(L4)   Sehgal [263] There exists a number $k, 0 < k < 1$, so that, for each $x, y \in X$,

$$d(Tx, Ty) \leq k \max\{d(x, y), d(x, Tx), d(y, Ty)\}$$

(L5)   Ćirić [99] There exists a number $k, 0 < k < 1$, so that, for each $x, y \in X$,

$$d(Tx, Ty) \leq k \max \left\{d(x, y), d(x, Tx), d(y, Ty), \frac{d(x, Ty) + d(y, Tx)}{2}\right\}$$

(L6)   [Chatterjea [89]] There is a $k \in [0, \frac{1}{2})$ so that

$$d(Tx, Ty) \leq k \left(d(x, Ty) + d(y, Tx)\right), \quad \forall x, y \in X.$$

(L6a) There are a $\alpha, \beta \in [0, 1)$ with $\alpha + \beta < 1$ so that

$$d(Tx, Ty) \leq \alpha d(x, Ty) + \beta d(y, Tx), \quad \forall x, y \in X.$$

(L7)   Reich [237] There are $\alpha, \beta, \gamma \in [0, 1)$ with $\alpha + \beta + \gamma < 1$ so that

$$d(Tx, Ty) \leq \alpha d(x, y) + \beta d(x, Ty) + \gamma d(y, Tx), \quad \forall x, y \in X.$$

(L7a) There are $\alpha, \beta, \gamma \in [0, 1)$ with $0 \leq \alpha + \beta < 1$ so that

$$d(Tx, Ty) \leq \alpha d(x, y) + \beta[d(x, Ty) + d(y, Tx)], \quad \forall x, y \in X.$$

(L8)   Ćirić [101] There are non-negative constants $\alpha, \beta, \gamma, \delta, \eta \in [0, 1)$ with $0 \leq \alpha + \beta + \gamma + 2\delta < 1$, so that, for each $x, y \in X$,

$$d(Tx, Ty) \leq \alpha d(x, y) + \beta d(x, Tx) + \gamma d(y, Ty) + \delta[d(x, Ty) + d(y, Tx)].$$

(L9)   [Hardy-Rogers [128]] There are non-negative constants $\alpha, \beta, \gamma, \delta, \eta$ with $0 \leq \alpha + \beta + \gamma + \delta + \eta < 1$, so that, for each $x, y \in X$,

$$d(Tx, Ty) \leq \alpha d(x, y) + \beta d(x, Tx) + \gamma d(y, Ty) + \delta d(x, Ty) + \eta d(y, Tx).$$

(L10) [Zamfirescu [280]] There are non-negative constants α, β, γ with $\alpha < 1, \beta < \frac{1}{2}$ and $\gamma < \frac{1}{2}$ so that at least one of the conditions is held:

a. $d(Tx, Ty) \leq \alpha d(x, y)$;
b. $d(Tx, Ty) \leq \beta[d(x, Ty) + d(y, Tx)]$;
c. $d(Tx, Ty) \leq \gamma[d(x, Tx) + d(y, Ty)]$,
   for each $x, y \in X$.

(L11) [Ćirić [100]] There exists a number $k, 0 \leq k < 1$, so that, for each $x, y \in X$,

$$d(Tx, Ty) \leq k \max\{d(x, y), d(x, Tx), d(x, Ty), d(y, Tx), d(y, Ty)\}.$$

Here, T is called quasi-contraction.

(L12) [Berinde [64]] There exists a number $k, 0 \leq k < 1$ and $L \geq 0$, so that, for each $x, y \in X$,

$$d(Tx, Ty) \leq kd(x, y) + Ld(x, Ty).$$

Here, T is called almost-contraction.

Theorem 3.8 *Let T be a self-mapping on a complete metric space (X, d). If one of the contractions in the linear form of (L1) to (L10) is fulfilled, then T has a unique fixed point.*

We shall use the following notations for orbits of x under the mapping T:

$$O(x, n) = \{x, Tx, \ldots, T^n x\}, \text{ for } n \in \mathbb{N},$$

$$O(x, \infty) = \{x, Tx, \ldots\}.$$

If each Cauchy sequence that is contained in $O(x, \infty)$, for some $x \in X$, converges in X, the we say that the metric space (X, d) is T-orbitally complete.

Theorem 3.9 *Let T be a self-mapping on a T-orbitally complete metric space (X, d). If T satisfies (L11) then, T has a unique fixed point.*

For the comparison of the contraction conditions which are discussed in Theorems 3.8 and 3.9, we refer to [240].

3.2.2 Nonlinear Extensions of Banach Contraction Principle

The first significant generalization of Banach's theorem was obtained by Rakotch [233], who replaced Banach's strict contractions by contractive mappings as follows.

Theorem 3.10 (Rakotch [233]) *Let (X, d) be a complete metric space and let T be a self-mapping on (X, d). If there is a monotone decreasing function $\alpha : (0, \infty) \to [0, 1)$ so that, for each $x, y \in X$,*

$$d(Tx, Ty) \leq \alpha(d(x, y))d(x, y), \tag{3.9}$$

then T possesses a unique fixed point x^*. In particular, for each $x \in X$, the sequence $\{T^n x\}$ converges to x^*.

Theorem 3.11 (Browder [84]) *Let (X, d) be a complete metric space and let Y be a bounded subset of X. Suppose that $T : Y \to Y$ satisfies*

$$d(Tx, Ty) \leq \phi(d(x, y)) \text{ for each } x, y \in Y,$$

where $\phi : [0, \infty) \to [0, \infty)$ is monotone nondecreasing and continuous from the right function so that $\phi(t) < t$ for each $t > 0$. then T has a unique fixed point x^*. In particular, for each $x \in X$, the sequence $\{T^n x\}$ converges to x^*. Further, if d_0 is the diameter of Y, then

$$d(T^n x, x^*) \leq \phi^n(d_0) \text{ and } \phi^n(d_0) \to 0 \text{ as } n \to \infty.$$

Theorem 3.12 (Boyd and Wong [80]) *Let T be a self-mapping on a complete metric space (X, d). We presume that there is an upper semi-continuous function from the right $\phi : [0, \infty) \to [0, \infty)$ with $\phi(t) < t$ for each $t > 0$. If T satisfies*

$$d(Tx, Ty) \leq \phi(d(x, y)) \text{ for each } x, y \in X,$$

then T has a unique fixed point x^*. In particular, for each $x \in X$, the sequence $\{T^n x\}$ converges to x^*.

Theorem 3.13 (Matkowski [209]) *Let T be a self-mapping on a complete metric space (X, d). We presume that there is a monotone nondecreasing function $\phi : (0, \infty) \to (0, \infty)$ with $\lim_{n \to \infty} \phi^n(t) = 0$ for each $t > 0$. If T satisfies*

$$d(Tx, Ty) \leq \phi(d(x, y)) \text{ for each } x, y \in X,$$

then T has a unique fixed point x^*. In particular, for each $x \in X$, the sequence $\{T^n x\}$ converges to x^*.

Proof Take an any $x \in X$ and let $x_n = T^n x$, $n \in \mathbb{N}$. We shall show that $\lim_{n \to \infty} d(x_n, x_{n+1}) = 0$.

$$0 \le \limsup_{n \to \infty} d(x_n, x_{n+1}) \le \limsup_{n \to \infty} \psi^n(d(x, x_1)) = 0.$$

Next, we demonstrate that the sequence $\{x_n\}$ is Cauchy.

By Lemma 2.4, we have $\psi(\varepsilon) < \varepsilon$ for each $\varepsilon > 0$. Since $\lim_{n \to \infty} d(x_n, x_{n+1}) = 0$, for given any $\varepsilon > 0$ it is possible to choose n so that

$$d(x_{n+1}, x_n) \le \varepsilon - \psi(\varepsilon).$$

Define

$$K(x_n, \varepsilon) = \{x \in X : d(x, x_n) \le \varepsilon\}.$$

Then if $z \in K(x_n, \varepsilon)$,

$$\begin{aligned} d(T(z), x_n) &\le d(f(z), T(x_n)) + d(T(x_n), x_n) \\ &\le \psi(d(z, x_n)) + d(x_{n+1}, x_n) \\ &\le \psi(\varepsilon) + (\varepsilon - \psi(\varepsilon)) = \varepsilon. \end{aligned}$$

Accordingly, we conclude that T is a self-mapping on $K(x_n, \varepsilon)$, that is, $T : K(x_n, \varepsilon) \to K(x_n, \varepsilon)$ and hence we have $d(x_m, x_n) \le \varepsilon$ for each $m \ge n$. Thus, we conclude that the sequence $\{x_n\}$ is Cauchy.

The conclusion of the proof follows by standard way.   □

Theorem 3.14 (Meir and Keeler [213]) *Let T be a self-mapping on a complete metric space (X, d). Suppose that for each $\varepsilon > 0$, there is $\delta > 0$ so that*

$$\varepsilon \le d(x, y) < \varepsilon + \delta \text{ implies } d(Tx, Ty) < \varepsilon,$$

for each $x, y \in X$. Then, for each $x \in X$, the sequence $\{T^n x\}$ converges to x^*. In addition, the limit x^* is the required unique fixed point of the selfmapping T.

Theorem 3.15 (Krasnosel'skiĭ and Zabrejko [199]) *Let T be a self-mapping on a complete metric space (X, d). For each $0 < \alpha < \beta$ there is $k(\alpha, \beta)$ so that*

$$\alpha \le d(x, y) \text{ implies } d(Tx, Ty) \le k(\alpha, \beta)d(x, y), \text{ for each } x, y \in X,$$

then, for each $x \in X$, the sequence $\{T^n x\}$ converges to x^*. In addition, the limit x^* is the required unique fixed point of the selfmapping T.

Theorem 3.16 (Burton [87]) *Let T be a self-mapping on a complete metric space (X, d). If for each $\lambda > 0$ there is $k(\lambda)$ so that*

$$d(x, y) \geq \lambda \text{ implies } d(Tx, Ty) \leq k(\lambda)d(x, y), \text{ for each } x, y \in X,$$

then, for each $x \in X$, the sequence $\{T^n x\}$ converges to x^*. In addition, the limit x^* is the required unique fixed point of the selfmapping T.

Theorem 3.17 (Geraghty [118]) *Let T be a self-mapping on a complete metric space (X, d). We presume that there is a mapping $\beta : (0, \infty) \to [0, 1)$ satisfying the condition,*

$$\beta(t_n) \to 1 \text{ implies } t_n \to 0.$$

If T fulfills
$$d(Tx, Ty) \leq \beta(d(x, y))d(x, y), \text{ for each } x, y \in X,$$

then, for each $x \in X$, the sequence $\{T^n x\}$ converges to x^*. In addition, the limit x^* is the required unique fixed point of the selfmapping T.

Theorem 3.18 *Let T be a self-mapping on a complete metric space (X, d). If one of the following assumptions $(N1)$ to $(N4)$ holds, then for each $x \in X$, the sequence $\{T^n x\}$ converges to x^*. In addition, the limit x^* is the required unique fixed point of the selfmapping T.*

$(N1)$ *[Das and Gupta [104]] There exist non-negative constants k_1, k_2 with $0 < k_1 + k_2 < 1$, so that, for each $x, y \in X$,*

$$d(Tx, Ty) \leq k_1 \frac{[1 + d(x, Tx)]d(y, Ty)}{1 + d(x, y)} + k_2 d(x, y).$$

$(N2)$ *[Khan [191], and independently, Fisher [114]] There exists a non-negative constants $k \in [0, 1)$ so that $x, y \in X$,*

$$\begin{aligned} d(Tx, Ty) &< \mu \frac{d(x, Tx)d(x, Ty) + d(y, Ty)d(y, Tx)}{d(x, Ty) + d(y, Tx)} && \text{if } d(x, Ty) + d(y, Tx) \neq 0, \\ d(Tx, Ty) &= 0, && \text{if } d(x, Ty) + d(y, Tx) = 0. \end{aligned}$$

$(N3)$ *[Alber and Guerre-Delabriere [10], and independently, Rhoades [241]] There exists a lower semi-continuous function $\varphi : [0, \infty) \to [0, \infty)$ so that*

$$d(Tx, Ty) \leq d(x, y) - \varphi(d(x, y)) \text{ for each } x, y \in X.$$

In the literature, the mapping T, satisfying the above inequality, is called weak-φ-contractive.

$(N4)$ *[Dutta and Choudhury [110]] There exist alternating distance functions $\vartheta, \varphi : [0, \infty) \to [0, \infty)$ so that*

$$\vartheta(d(Tx, Ty)) \leq \vartheta(d(x, y)) - \varphi(d(x, y)) \text{ for each } x, y \in X.$$

In the literature, the mapping T, satisfying the above inequality, is called weak-$\vartheta - \varphi$-contractive.

Note that it is possible to put more assumptions into the above theorem. On the other hand, we prefer to skip these further possible assumptions in the above theorem since most of them are slightly modified versions of listed assumptions. In fact, several extension in metric fixed point theory is either equivalent to or subsequent to the existing results, see e.g. [8, 25, 26, 107, 108, 126, 139–141, 150, 158, 234, 244, 256, 259]

Theorem 3.19 (Caristi [90])   *Let T be a continuous self-mapping on a complete metric space (X, d). If there is a function $\varphi : X \to [0, \infty)$, so that*

$$d(x, Tx) \leq \varphi(x) - \varphi(Tx), \qquad x \in X, \tag{3.10}$$

then, for each $x \in X$, the sequence $\{T^n x\}$ converges to x^*. In addition, the limit x^* is the required unique fixed point of the selfmapping T.

Proof This idea is already used in the alternative proof of Banach's contraction principle. The inequality (3.10) yields that $\{\varphi(T^n x)\}$ is monotone decreasing and bounded below. Accordingly, we conclude that $\lim_{n\to\infty} \varphi(T^n(x)) = L \geq 0$. On account of the triangle inequality, for $m, n \in \mathbb{N}$ with $m > n$, we have

$$d(T^n x, T^m x) \leq \sum_{i=n}^{m-1} d(T^i x, T^{i+1} x) \leq \varphi(T^n x) - \varphi(T^m x).$$

Thus, we get $\lim_{m,n\to\infty} d(T^n x, T^m x) = 0$. Since X is complete there is $x^* \in X$ so that $\lim_{n\to\infty} T^n x = x^*$ and by continuity of T, we find $x^* = Tx^*$. □

Notice that from Theorem 3.19, it is possible to obtain an estimate on the rate of convergence of $\{T^n x\}$ by regarding the inequality

$$\sum_{i=n}^{m-1} d(T^i x, T^{i+1} x) \leq \varphi(T^n x) - \varphi(T^m x).$$

It yields

$$d(T^n x, T^m x) \leq \varphi(T^n x) - \varphi(T^m x),$$

and if $x^* = Tx^*$, upon letting $m \to \infty$ one has

$$d(T^n x, x^*) \leq \varphi(T^n x) - \varphi\left(x^*\right).$$

3.3   Locally Contractive Mappings and Related Fixed Point Results

On a metric space (X, d), a mapping T from X to itself is locally contractive if for every $x \in X$ there exist ε and k $(\varepsilon > 0, 0 < k < 1)$, which may depend on p, so that:

$$x, y \in S(p, \varepsilon) = \{q : 0 < d(p, q) < \varepsilon\} \text{ implies (3.2)} \qquad (3.11)$$

A mapping T of X into itself is said to be (ε, k)-uniformly locally contractive if it is locally contractive and both k and ε do not depend on p. In this aspect, a mapping T satisfying (3.2) is called globally contractive mapping. Under this setting, a globally contractive mapping can be regarded as a (∞, k) uniformly locally contractive mapping. A metric space (X, d) is called ε-chainable if for every $a, b \in X$ there is an ε-chain, that is a finite set of points $a = x_0, x_1, \ldots, x_n = b$ (n may depend on both a and b) so that $d(x_{i-1}, x_i) < \varepsilon$ for $i = 1, 2, \ldots, n$.

Theorem 3.20 (Edelstein [111]) *Let T be a continuous self-mapping on a complete ε-chainable metric space (X, d). If it is (ε, k)-uniformly locally contractive then there is a unique point $x^* \in X$ so that $T x^* = x^*$.*

Proof For an arbitrary point $x \in X$, we consider the ε-chain as follows:

$$x := x_0, x_1, \ldots, x_m := Tx.$$

By the triangle inequality, we derived that

$$d(x, Tx) = d(x_0, x_m) \leq \sum_{i=1}^{m} d(x_{i-1}, x_i) < m\varepsilon. \qquad (3.12)$$

On the other hand, since T is (ε, k)-uniformly locally contractive, we have

$$d(T x_{i-1}, T x_i) < k d(x_{i-1}, x_i) < k\varepsilon.$$

Iteratively, we find that

$$d(T^n x_{i-1}, T^n x_i) < k d(T^{n-1} x_{i-1}, T^{n-1} x_i) < \cdots < k^n \varepsilon. \qquad (3.13)$$

Consequently, by combining (3.12) and (3.13) we get that

$$d(T^j x, T^{j+1} x) = d(T^j x, T^j(Tx)) \leq \sum_{i=1}^{m} d(T^j x_{i-1}, T^j x_i) < k^n m\varepsilon.$$

We shall show that the sequence $\{T^n x\}$ is Cauchy. Indeed,

$$d(T^n x, T^{n+r} x) \le \sum_{j=n}^{n+r-1} d(T^j x, T^{j+1} x) < \frac{k^n}{1-k} m\varepsilon \to 0 \text{ as } n \to \infty.$$

Due to completeness of the metric space, there is $x^* \in X$ so that $\lim_{n \to \infty} T^n x = x^*$. By continuity, we have $Tx^* = x^*$. For the uniqueness, we assume, on the contrary, that there is $y^* \in X$ with $Ty^* = y^*$ and $y^* \ne x^*$, that is $d(x^*, y^*) > 0$. Since it is ε-chainable space, we have

$$x^* = x_1, x_2, \ldots, x_q = y^*.$$

Thus, we have

$$d(x^*, y^*) = d(T^p x^*, T^p y^*) \le \sum_{i=1}^{q} d(T^p x_{i-1}, T^p x_i) < k^q p\varepsilon \to 0 \text{ as } q \to \infty,$$

a contradiction. Hence, x^* is the unique fixed point of T.   □

3.4   Discussion on the Mappings Whose Iteration form a Contraction

Although it was not mentioned in Banach's contraction mapping principle, the continuity of the mapping is necessarily continuous. On the other hand, we can not model every problem in the real world with continuous functions. Indeed, the requirement of continuity for the considered map is a massive condition, and a vast number of the functions derived from real-world problems are discontinuous. At this point, a quite natural question arises: Whether discontinuous mapping possesses a fixed point that fulfills a kind of contraction? This question was responded affirmatively by Bryant [85] with the following plain but meaningful example:

Example

[cf. [85]] Let T be a selfmapping on $[0, 3]$ so that

$$T(x) = \begin{cases} 0 & \text{if } x \in [0, 2], \\ 2 & \text{if } x \in (2, 3]. \end{cases}$$

It is evident that the selfmapping T is discontinuous at 2. Consequently, T does not form a contraction. However, the 2nd iteration of T is equal to 0 for each $x \in [0, 3]$. Accordingly, T^2 is a contraction on $[0, 3]$ with a fixed point 0.

Based on the discussion in the previous example, a relaxed version of the Banach-Caccioppoli theorem was proposed by Bryant in the following way:

Theorem 3.21 (Bryant Theorem [85]) *Let T be a self-mapping on a complete metric space* (X, d). *If there is* $k \in [0, 1)$ *so that*

$$d(T^m v, T^m w) \leq kd(x, y), \tag{3.14}$$

for each $x, y \in X$, *where* $m \in \mathbb{N}$, *then T possesses a unique fixed point.*

Proof By taking, Theorem 3.2, into account, we derive that T^m (that is, m-th iteration of T) possesses a unique fixed point, say $x^* \in X$. Employing the elementary equality below,

$$Tx^* = T(T^m x^*) = T^m(Tx^*),$$

we deduce that the Tx^* is also fixed point of T^m. Due to the uniqueness of the limit, we necessarily have $Tx^* = x^*$.   □

We underline the fact that the condition (3.14) does not force T be continuous but its m-th iteration. Notice also that the mapping satisfying the condition (3.14) are Bessage operator.
    Now, we consider another example to illustrate the idea of Bryant:

Example

Let $X = [0, \infty)$, $d : X \times X \to X \to [0, \infty)$, $d(x, y) = |x - y|$ and $T : X \to X$,

$$Tx = \begin{cases} 2, & \text{if } x \in [0, 2] \\ \frac{1}{2}, & \text{if } x \in (2, 4] \\ \frac{4x^2+1}{x^2+4x}, & \text{if } x \in (4, \infty) \end{cases}$$

We have

$$T^2 x = \begin{cases} 2, & \text{if } x \in [0, 4] \\ \frac{1}{2}, & \text{if } x \in (4, \infty) \end{cases}$$

and

$$T^3 x = 2.$$

Therefore

1. $d(T2, T\frac{5}{2}) = d(2, \frac{1}{2}) = \frac{3}{2} > k \cdot \frac{1}{2} > k \cdot d(2, \frac{5}{2})$;
2. $d(T^2 4, T^2 \frac{9}{2}) = d(2, \frac{1}{2}) = \frac{3}{2} > k \cdot \frac{1}{2} > k \cdot d(4, \frac{9}{2})$ any $x \in [0, \infty)$ and   then
   $d(T^2 x, T^2 y) = 0 < k \cdot d(x, y)$.
3. On the other hand,
$$d(T^3 x, T^3 y) = 0 < k \cdot d(x, y),$$

for each $x, y \in X$.

Notice that 2 is the fixed point of T, T^2 and T^3, simultaneously.

Although Banach's fixed theorem is necessarily continuous, in the main theorem of Bryant [85] the continuity of the mapping is not required. On the other hand, it led to the initiation of contractive type fixed point results introduced by Sehgal [264]. In this new trend, the theorems are based on pointwise considerations and constructions. The first renowned generalization of Bryant's theorem was proposed by Sehgal [264] in 1969.

Theorem 3.22 (Sehgal Theorem [264]) *We presume that the non-empty set X equipped with the metric d constitutes a complete metric space and $T : X \to X$ is continuous selfmapping. If there is a $k \in [0, 1)$ so that for each $x \in X$, there is a positive integer $n(x)$ with*

$$d(T^{n(x)}(x), T^{n(x)}(y)) \leq kd(x, y), \tag{3.15}$$

for each $y \in X$, then, for each $x \in X$, the sequence $\{T^n x\}$ converges to x^*. In addition, the limit x^* is the required unique fixed point of the selfmapping T.

We say that a mapping $T : X \to X$ is with a contractive iterate at a point x if it satisfies (3.15).

Example [264]

Let $A_n = \left[\dfrac{1}{2^n}, \dfrac{1}{2^{n-1}} \right]$ for each $n \in \mathbb{N}$ and $X = [0, 1] = \{0\} \cup \bigcup\limits_{n=1}^{\infty} A_n$ equipped with the usual metric $d(x, y) := |x - y|$ for each $x, y \in X$. For each $n \in \mathbb{N}$, we define $T : A_n \to A_{n+1}$ as

$$T(x) = \begin{cases} 0 & \text{if } x = 0; \\ \dfrac{n+2}{n+3}\left(x - \dfrac{1}{2^{n-1}}\right) + \dfrac{1}{2^n} & \text{if } x \in \left[\dfrac{3n+5}{2^{n+1}(n+2)}, \dfrac{1}{2^{n-1}} \right]; \\ \dfrac{1}{2^{n+1}} & \text{if } x \in \left[\dfrac{1}{2^{n-1}}, \dfrac{3n+5}{2^{n+1}(n+2)} \right] \end{cases}$$

Observe that the function T, defined on X, is a continuous, non-decreasing. Further, 0 is the unique fixed point of T. In fact, setting $k = \frac{1}{2}$, for each $x \in A_n$, the integer $n(x)$ can be chosen as $n + 3$ where $n(0) \geq 1$. On the other hand, the mapping T does not form a contraction. In addition, the condition (3.15) is weaker then (3.14). See [264] for detailed discussion on it.

In 1970, Guseman [122] refined the statement of Sehgal [264] by removing the superflu-ous condition: "continuity of the mapping". As Bryant [85], Guseman [122] didn't assume the "continuity" of the mentioned mapping.

Theorem 3.23 (Guseman Theorem [122]) *Let T be a self-mapping on a complete metric space (X, d). We presume that there is $B \subset X$ in a way that*

(a) $T(B) \subset B$,
(b) *for some $k < 1$ and each $y \subset B$ there is an integer $n(y) \geq 1$ with*

$$d(T^{n(y)}(x), T^{n(y)}(y)) \leq kd(x, y) \text{ for each } x \subset B,$$

(c) *for some $x_0 \subset B$, $cl\{T^n(x_0) : n \geq 1\} \subset B$.*

Then, for each $y_0 \in B$, we have $T^n(y_0) \to x^*$. In addition, x^* is the unique fixed point of T. In addition, if $d(T^{n(x^*)}(x), T^{n(x^*)}(x^*)) \leq kd(x, x^*)$ for each $x \in X$, then $x^* \in X$ is unique fixed point and $T^n(x_0) \to x^*$ for each $x_0 \in X$.

Theorem 3.24 (Guseman Uniqueness Theorem [122]) *Let T be a self-mapping on a com-plete metric space (X, d). We presume that $x^*, x_0 \in X$ so that $T^n(x_0) \to x^*$. If T is with a contractive at a point x^*, then T has a unique fixed point $x^* \in X$ and $T^n(y) \to x^*$ for each $y \in X$.*

Theorem 3.25 (Matkowski Theorem [212]) *We presume that T is a self-mapping on a complete metric space (X, d). We also assume that an auxiliary function $\alpha : [0, \infty)^5 \to [0, \infty)$ satisfies $\gamma(t) = \alpha(t, t, t, 2t, 2t)$ for $t \geq 0$. Assume also that*

(i) $\lim_{n \to \infty} \gamma^n(t) = 0, t > 0$,
(ii) $\lim_{t \to \infty} (t - \gamma(t)) = \infty$,
(iii) *α is nondecreasing with respect to each variable,*
(iv) *for each $x \in X$, there is $n(x) \in \mathbb{N}$ so that for each $y \in X$,*

$$d(T^{n(x)}(x), T^{n(x)}(y)) \leq \Gamma_T^n(x, y), \tag{3.16}$$

where

$$\Gamma_T^n(x, y) = \alpha\left(d(x, T^{n(x)}(x)), d(x, T^{n(x)}(y)), d(x, y), d(T^{n(x)}(x), y), d(T^{n(x)}(y), y)\right).$$

Then, the sequence $\{T^n x\}$ converges to x^*, for each $x \in X$. In addition, the limit x^* is the required unique fixed point of the selfmapping T.

Lemma 3.1 ([212]) *We presume that the function $\psi : [0, \infty) \to [0, \infty)$ is non-decreasing so that $\lim_{n \to \infty} \psi^n(t) = 0$ for $t > 0$. Then, we find $\psi(t) < t$.*

In the following theorem, the common fixed point for contractive type mappings were discussed by Ray and Rhoades [236] in 1977.

Theorem 3.26 (Ray and Rhoades [236]) *Let T and S be self-mappings on a complete metric space (X, d). If there is a real number $k(0 < k < 1)$ so that there are $n(x), m(y) \in \mathbb{N}$, so that*

$$d(T^{n(x)}x, S^{m(y)}y) \leq k \max \left\{ d(x, y), d(x, T^{n(x)}x), d(y, S^{m(y)}y), \frac{d(x, S^{m(y)}y) + d(y, T^{n(x)}x)}{2} \right\},$$
(3.17)

for each $x, y \in X$. Then S and T have a unique common fixed point.

In 1979, Singh [267] reported the following result:

Theorem 3.27 (Singh [267]) *We presume that T is a self-mapping on a complete metric space (X, d). We also assume that for each $x \in X$, there is $n(x) \in \mathbb{N}$ so that for each $y \in X$*

$$d(T^{n(x)}(x), T^{n(x)}(y)) \leq q(x, y)d(x, y) + r(x, y)d(x, T^{n(x)}(x)) + s(x, y)d(y, T^{n(x)}(y))$$
$$+t(x, y)d(x, T^{n(x)}(y)) + m(x, y)d(y, T^{n(x)}(x))$$
(3.18)

where $q(x, y), r(x, y), s(x, y), t(x, y)$, and $m(x, y)$ are positive-valued functions so that $\sup_{x,y \in X} \{2(m(x, y) + s(x, y) + t(x, y)) + q(x, y + r(x, y))\} = \lambda < 1$. Then, for each $x_0 \in X$, we have $T^n(x_0) \to x^*$. In addition, the limit x^* is the required unique fixed point of the selfmapping T.

The following significant theorem was proved by Ćirić [96] in 1983.

Theorem 3.28 (Ćirić [96]) *We presume that T is a self-mapping on a complete metric space (X, d). If there is a real number $k(0 < k < 1)$ so that for each $x \in X$ there is $n = n(x) \in \mathbb{N}$ so that for each $y \in X$,*

$$d(T^n x, T^n y) \leq k \cdot \max \left\{ d(x, y), d(x, Ty), ..., d(x, T^n y), d(x, T^n x) \right\}.$$
(3.19)

Then, for each $x \in X$, the sequence $\{T^n x\}$ converges to x^*. In addition, the limit x^* is the required unique fixed point of the selfmapping T.

In 1990, Kincses-Totik [196] proved the following.

Theorem 3.29 (Kincses-Totik [196, Theorem 3]) *Let T be a continuous selfmapping on a compact metric space (X, d). If for each $x \in X$ there is $p(x) \in \mathbb{N}$ so that for every $y \in X$, $y \neq x$ we have*

$$d(T^{n(x)}(x), T^{n(x)}(y)) < \max \left\{ d(x, T^{n(x)}(y)), d(y, T^{n(x)}(x)), d(y, x) \right\}, \tag{3.20}$$

then, for each $x \in X$, the sequence $\{T^n x\}$ converges to x^*. In addition, the limit x^* is the required unique fixed point of the selfmapping T.

Kincses-Totik [196] proved the followings:

Theorem 3.30 (Kincses-Totik) *We presume that T is a self-mapping on a complete metric space (X, d). If $\alpha < 1$ and to every $x \in X$ there is $p(x) \in \mathbb{N}$ so that for every $y \in X$*

$$d(T^{n(x)}(x), T^{n(x)}(y)) \leq \alpha \max \left\{ d(x, T^{n(x)}(y)), d(y, T^{n(x)}(x)), d(y, x) \right\}, \tag{3.21}$$

for each $x \in X$, then the sequence $\{T^n x\}$ converges to x^*. In addition, the limit x^* is the required unique fixed point of the selfmapping T.

Theorem 3.31 (Kincses-Totik) *We presume that T is a self-mapping on a complete metric space (X, d). We also assume that there is a monotone decreasing function $\alpha : (0, \infty) \to [0, 1)$ and for each $x \in X$, there is $n(x) \in \mathbb{N}$ so that for each $y \in X$ $y \neq x$ we have*

$$d(T^{n(x)}(y), T^{n(x)}(x)) < \alpha(d(y, x)). \tag{3.22}$$

Then, for each $x \in X$, the sequence $\{T^n x\}$ converges to x^*. In addition, the limit x^* is the required unique fixed point of the selfmapping T.

In 1995, Jachymski [138] gave the latest result in the setting of a complete metric space:

Theorem 3.32 ([138]) *Let T be a self-mapping on a complete metric space (X, d). Suppose that an auxiliary mapping $p : X \to \mathbb{N}$ satisfies*

(i) *nondecreasing;*
(ii) *$\psi(t) < t$ for each $t > 0$;*
(iii) *$\lim_{t \to \infty} (t - \psi(t)) = \infty$.*

If for so that for each $x, y \in X$

$$d(T^{p(x)}x, T^{p(x)}y) \leq \psi \left[\max \left\{ d(x, T^j y), d(T^{p(x)}x, T^j y), j = 0, 1, 2, ...p(x) \right\} \right], \tag{3.23}$$

then, T possesses s fixed point x^*.

Corollary 3.1 ([138]) *Let T be a selfmap of a complete metric space (X, d) and let $p : X \to \mathbb{N}$ so that for each $x, y \in X$*

$$d(T^{p(x)}x, T^{p(x)}y) \le I(x, y), \tag{3.24}$$

where

$$I(x, y) := \psi \left[\max \left\{ d(x, T^{p(x)}(y)), d(y, T^{p(x)}(x)), d(y, x) \right\} \right]$$

and ψ is a nondecreasing function so that $\psi(t) < t$ for each $t > 0$ and $\lim_{t \to \infty}(t - \psi(t)) = \infty$. Then T possesses s fixed point x^*.

Corollary 3.2 ([138]) *Let T be a selfmap of a complete metric space (X, d) and let $p : X \to \mathbb{N}$ so that for each $x, y \in X$*

$$d(T^{p(x)}x, T^{p(x)}y) \le J(x, y), \tag{3.25}$$

where $a, b, c \in \mathbb{R}_+$ and $4a + 2b + c < 1$ and

$$J(x, y) := a[d(x, T^{p(x)}(x)) + d(y, T^{p(x)}(y))] + b[d(x, T^{p(x)}(y)) + d(y, T^{p(x)}(x))] + cd(y, x).$$

Then T possesses s fixed point x^*.

3.5   Extension on Compact Metric Spaces

In this section, we consider the contractive mappings and related fixed point results of Nemytskii [220] and Edelstein [112]. In 1936, Nemytskii [220] proved that a contractive mapping in a compact metric space possesses a fixed point. First, we shall provide an example to clarify why a metric's "completeness" is insufficient and why we need the "compactness" of a metric on which the mentioned contractive mapping is defined.

Example

[See e.g. [98]] Let $X = [1, \infty)$ equipped with a usual metric $d(x, y) = |x - y|$. Suppose that a selfmapping T from X into itself is defined by $Tx = x + \frac{1}{x}$. It is clear that T is a contractive mapping. Indeed, for each distinct pair $x, y \in X$, we observe that

$$d(Tx, Ty) = |x + \tfrac{1}{x} - y - \tfrac{1}{y}| = |x - y + \tfrac{1}{x} - \tfrac{1}{y}|$$
$$= \left(1 - \tfrac{1}{xy}\right)|x - y| = \left(1 - \tfrac{1}{xy}\right)$$
$$< d(x, y),$$

since $\left(1 - \frac{1}{xy}\right) < 1$. On the other hand T is fixed point free, that is, fixed point equation $x = Tx = x + \frac{1}{x}$ has no solution.

Theorem 3.33 (Nemytskii [220]) *On a compact metric space* (X, d) *each contractive self-mapping of X, that is,*

$$d(Tx, Ty) < d(x, y) \text{ for each distinct } x, y \in X, \tag{3.26}$$

possess a unique fixed point.

Proof We, first, define a function $\theta(x) = d(x, Tx)$. On account the fact that T is necessarily continuous, the function θ becomes continuous either. On account of X is compact, there is a point $x^* \in X$ so that

$$\theta(x^*) = d(x^*, Tx^*) = \min_{x \in X}\{d(x, Tx)\}.$$

If $x^* = Tx^*$, then, the proof is completed. Suppose, on the contrary, that $Tx^* = x^*$. Since T is contractive, we have

$$\theta(Tx^*) = d(Tx^*, TTx^*) < d(x^*, Tx^*) = \theta(x^*),$$

which contradicts with the fact that x^* is the minimum. Consequently, x^* is a fixed point of T. Uniqueness follows from the definition of mapping (3.26).   □

Theorem 3.34 (Hardy-Rogers [128]) *Let* (X, d) *be a compact metric space and T be a selfmapping on X. If there are non-negative constants* $\alpha, \beta, \gamma, \delta, \eta$ *with* $0 \leq \alpha + \beta + \gamma + \delta + \eta < 1$, *so that, for each distinct* $x, y \in X$,

$$d(Tx, Ty) < \alpha d(x, y) + \beta d(x, Tx) + \gamma d(y, Ty) + \delta d(x, Ty) + \eta d(y, Tx),$$

then T possess a unique fixed point.

Theorem 3.35 (Edelstein [112]) *Let* (X, d) *be a metric space and T be a contractive selfmapping on X. If there is* $x_0 \in X$ *so that the its iterative sequence* $\{T^n x_0\}$ *contains a convergent subsequence* $\{T^{n_k} x_0\}$, *then* $x^* = \lim_{k \to \infty} T^{n_k} x_0$ *is a unique fixed point of T.*

Instead of given the original proof of Edelstein [112], we prefer to state the simpler one which is belong to Ćirić [98].

Proof We construct an iterative sequence for the given initial value x_0 so that $x_{n+1} := Tx_n$ for each $n \in \mathbb{N}_0$. We assume that $x_n \neq x_{n+1}$ for each $n \in \mathbb{N}_0$. Indeed, if there is n_0 so that $x_{n_0} \neq x_{n_0+1} = Tx_{n_0}$, then x_{n_0} forms a fixed point that terminates the proof.

Employing the contractive condition (3.26), by substituting $x = x_n$ and $y = x_{n+1}$, we find

$$d(x_{n+1}, x_{n+2}) = d(Tx_n, Tx_{n+1}) < d(x_n, x_{n+1}),$$

for any $n \in \mathbb{N}$. Accordingly, we deduce that the sequence $\{d(x_n, x_{n+1})\}$ is positive and strictly decreasing. Hence, the sequence $\{d(x_{n+1}, x_n)\}$ is convergent to d^*.

On the other hand, by assumption of the theorem, the iterative sequence $\{x_n\} = \{T^n x_0\}$ has a convergent subsequence $\{x_{n_k}\}$ so that $x^* = \lim_{k \to \infty} x_{n_k}$. Since T is necessarily continuous, we derive that

$$T x^* = \lim_{k \to \infty} T x_{n_k} = \lim_{k \to \infty} T x_{n_k+1} \text{ and also } T^2 x^* = \lim_{k \to \infty} T x_{n_k+2}.$$

Consequently, regarding the continuity of the metric d, together with the observation above, we have

$$d(x^*, T x^*) = \lim_{k \to \infty} d(x_{n_k}, x_{n_k+1}),$$
$$d(T x^*, T^2 x^*) = \lim_{k \to \infty} d(x_{n_k+1}, x_{n_k+2}).$$

Since the sequences $\{d(x_{n_k}, x_{n_k+1})\}$ and $\{d(x_{n_k+1}, x_{n_k+2})\}$ are subsequences, they have the same limit d^*, that is,

$$d^* = d(x^*, T x^*) = d(T x^*, T^2 x^*).$$

We assert that $T x^* = x^*$. Suppose, on the contrary, $T x^* \neq x^*$. Then, we can apply (3.26) that yields

$$d(T x^*, T^2 x^*) < d(x^*, T x^*),$$

a contradiction. Thus, x^* is the required fixed point. The uniqueness of the fixed point is derived from (3.26) easily.   □

Notice that in a compact metric space, every sequence has a convergent subsequence. Hence, the result of Nemytskii [220], Theorem 3.33, is covered by the corresponding results of Edelstein [112], Theorem 3.35.

Definition 3.3 (*Bailey* [55]) x is proximal to y under T provided for each $\varepsilon > 0$ there is $n \in \mathbb{N}$ so that $d(T^n x, T^n y) < \varepsilon$. If x and y are not proximal they are said to be distal. If for each $\varepsilon > 0$, there is $n \in \mathbb{N}$ so that

$$d(T^m x, T^m y) < \varepsilon,$$

for each $m \geq n$, then x and y are said to be asymptotic under T.

Notice that the condition $x \neq y$ is not required.

Lemma 3.2 (Bailey [55]) *Suppose that T is a continuous selfmapping on a compact metric space (X, d). If x is proximal to $T^p x$ under T for some $p \in \mathbb{N}$ and $x \in X$, then there is $z \in X$ of period not more than p.*

Proof Set $\varepsilon = \frac{1}{k}$ and strictly increasing sequence $\{n_k\}$ so that

$$d(T^{n_k}x, T^{n_k}(T^p x)) < \frac{1}{k}.$$

On account of compactness, we suppose that $T^{n_k}x \to z$, and $T^{n_k}(T^p x) \to w$, as $k \to \infty$, for some $z, w \in X$. Observe that $z = w$. On the other hand, by continuity of T, we have $T^k(z) = w$. Hence, $T^k(z) = w = z$ that is, z is a periodic point at most k. $\quad\square$

A selfmapping mapping T on a metric space (X, d) is called ε-contractive or local contractive if the following condition is fulfilled:

(LC) There exists $\varepsilon > 0$ so that $0 < d(x, y) < \varepsilon$ implies $d(Tx, Ty) < d(x, y)$.

On the other hand, we say that T is iteratively local contractive if

(ILC) $0 < d(x, y)$ implies there is a positive integer $n = n(x, y)$ so that $d(T^n x, T^n y) < d(x, y)$.

Furthermore a selfmapping mapping T on a metric space (X, d) is called eventually contractive if the following expression holds

(EC) there is $\varepsilon > 0$ so that $0 < d(x, y) < \varepsilon$ implies $d(T^n x, T^n y) < d(x, y)$.

Theorem 3.36 (Bailey [55]) *Suppose that T is a continuous selfmapping on a compact metric space (X, d). If T is iteratively local contractive (that is, T satisfies (ILC)), then every pair of points in X is proximal under T.*

Proof Take an arbitrary pair of points $x, y \in X$. If there is $n_0 \in \mathbb{N}$ so that $T^{n_0}x = T^{n_0}y$, then x and y are proximal. This completes the proof. Accordingly, we suppose the converse, that is, $d(T^n x, T^n y) > 0$ for each $n \in \mathbb{N}$. Since T is iteratively local contractive, we can construct a sequence $\{n_i\}$ in a way that the condition

$$\cdots < d(T^{n_k}x, T^{n_k}y) < \cdots < d(T^{n_2}x, T^{n_2}y) < d(T^{n_1}x, T^{n_1}y) < d(x, y)$$

is satisfied where each n_i is chosen as small as possible. Observe that for $k \le n_i$, we have

$$d(T^{n_i}x, T^{n_i}y) < d(T^k x, T^k y).$$

If x and y are proximal, then there is nothing to prove. Attendantly, we suppose that x and y are distal and hence there is a $\delta > 0$ so that $d(T^n x, T^n y) \ge \delta$ for each $n \in \mathbb{N}$. On account of compactness, we may assume that $T^{n_i}x \to z$ and $T^{n_i}y \to w$, as $k \to \infty$, for some $z, w \in X$. Note that $z \ne w$ since x and y are distal. On the other hand, $n_i + k \le n_{i+k}$

and hence
$$d(T^k zx, T^k w) = \lim_{i \to \infty} d(T^k(T^{n_i} x), T^k(T^{n_i} y))$$
$$= \lim_{i \to \infty} d(T^{n_i+k} x, T^{n_i+k} y)$$
$$\geq \lim_{i \to \infty} d(T^{n_i+k} x, T^{n_i+k} y) = d(z, w).$$

for each $k \in \mathbb{N}$. It contradict the assumption that T is iteratively local contractive. As a result, any pairs of points in X is proximal under T. $\square$

Corollary 3.3 *On a compact metric space* (X, d), *if a self-mapping* $T : X \to X$ *is contractive (3.26), then every pair of points is asymptotic under* T.

Sketch of the Proof On account of Theorem 3.33, the mapping T has a unique fixed point x^*, and, moreover, each sequence $\{T^n x\}$ converges to x^* for each x. On the other hand, T is necessarily continuous due to (3.26). By employing Theorem 3.36, we deduce that every pair of points is asymptotic under T.

Corollary 3.4 *On a compact metric space* (X, d), *if a self-mapping* $T : X \to X$ *satisfies* (ILC), *then* T *possess a unique fixed point in* X.

Proof On account of Theorem 3.36, we deduce that every pair of points in X is proximal under T. In particular, x^* and Tx^* are proximal. Taking Lemma 3.2 into account, we conclude that x^* is the unique fixed point of T. $\square$

Corollary 3.5 *On a compact metric space* (X, d), *if a self-mapping* $T : X \to X$ *satisfies* (ILC), *then for each* $k \in \mathbb{N}$, *the k-th iterate of* T, T^k, *satisfies* (ILC) *either.*

Proof Suppose, on the contrary that T satisfies (ILC), but not T^k. Thus, there exist $k \in \mathbb{N}$, and a pair x, y so that

$$0 < d(x, y) \leq d((T^k)^n x, (T^k)^n) = d(T^{kn} x, T^{kn} y), \text{ for each } n \in \mathbb{N}.$$

Thus, there is an $\varepsilon > 0$ so that $d(T^m x, T^m y) > \varepsilon$ for each $m \in \mathbb{N}$ and hence the pair of x and y is distal. This contradicts Theorem 3.36. $\square$

Theorem 3.37 (Bailey [55]) *Suppose that* T *is a continuous selfmapping on a compact metric space* (X, d). *If* T *is eventually l-contractive (that is,* T *satisfies the hypothesis* (EC)*), then every pair of points in* X *is proximal under* T.

Proof Take any point $x, y \in X$ so that $d(x, y) < \varepsilon$. After then, employ the proof of Theorem 3.36. $\square$

3.6   Nonunique Fixed Point Theorems

The notion of nonunique fixed point was proposed by Ćirić [97]. After this interesting pioneer result, many researchers have studied on this directions and reported several attractive results, see e.g. [3, 4, 101, 142, 153, 156, 159, 167, 223]. Throughout this section, we presume that metric space (X, d) is orbitally complete and T be a selfmapping on X.

Theorem 3.38   (Ćirić's nonunique fixed point theorem [97]) *We presume that there is a real number $k \in [0, 1)$ so that*

$$\min\{d(y, Ty), d(x, Tx), d(Tx, Ty)\} - \min\{d(Tx, y), d(x, Ty)\} \leq kd(x, y),$$

for each $x, y \in X$. Then, for each $x \in X$, the sequence $\{T^n x\}$ converges to x^*. In addition, the limit x^* is the required fixed point of the selfmapping T.

Theorem 3.39   (Achari's nonunique fixed point [4]) *We presume that there is a real number $k \in [0, 1)$ so that for each $x, y \in X$,*

$$\frac{A(x,y) - B(x,y)}{C(x,y)} \leq kd(x, y), \tag{3.27}$$

where

$$C(x, y) = \min\{d(x, Tx), d(y, Ty)\},$$
$$B(x, y) = \min\{d(x, Tx)d(x, Ty), d(y, Ty)d(Tx, y)\},$$
$$A(x, y) = \min\{d(Tx, Ty)d(x, y), d(x, Tx)d(y, Ty)\},$$

with $C(x, y) \neq 0$. Then, for each $x \in X$, the sequence $\{T^n x\}$ converges to x^*. In addition, the limit x^* is the required fixed point of the selfmapping T.

Theorem 3.40   (Pachpatte's Nonunique fixed point [223]) *We presume there is a real number $k \in [0, 1)$ so that*

$$m(x, y) - n(x, y) \leq kd(x, Tx)d(y, Ty), \tag{3.28}$$

for each $x, y \in X$, where

$$n(x, y) = \min\{d(x, Tx)d(y, Ty), d(x, Ty)d(y, Tx)\},$$
$$m(x, y) = \min\{[d(Tx, Ty)]^2, d(x, y)d(Tx, Ty), [d(y, Ty)]^2\}.$$

Accordingly, T possesses a fixed point in X. More precisely, the recursive sequence $\{T^n x_0\}_{n \in \mathbb{N}}$ converges to a fixed point of T, for a random initial point $x_0 \in X$.

Theorem 3.41   (Ćirić-Jotić's nonunique fixed point [101]) *We presume that there are real numbers $a \geq 0$ and $k \in [0, 1)$ so that*

$$J(x, y) - aI(x, y) \leq kL(x, y), \tag{3.29}$$

for each distinct $x, y \in X$ where

$$L(x, y) = \max\{d(x, y), d(x, Tx)\},$$

$$J(x, y) = \min \left\{ \begin{array}{l} d(Tx, Ty), d(x, y), d(x, Tx), d(y, Ty), \frac{d(x,Tx)[1+d(y,Ty)]}{1+d(x,y)}, \\ \frac{d(y,Ty)[1+d(x,Tx)]}{1+d(x,y)}, \frac{\min\{d^2(Tx,Ty),d^2(x,Tx),d^2(y,Ty)\}}{d(x,y)} \end{array} \right\},$$

$$I(x, y) = \min\{d(y, Tx), d(x, Ty)\}.$$

Then, for each $x \in X$, we deduce that $\{T^n x\}$ converges to x^*. In addition, the limit x^* is the required fixed point of the selfmapping T.

Theorem 3.42 (Karapınar's nonunique fixed point [156]) *We presume that there are non-negative real numbers a_1, a_2, a_3, a_4, a_5 and a self mapping $T : X \to X$ satisfies the conditions*

$$0 \le \frac{a_4 - a_2}{a_1 + a_2} < 1, \quad a_1 + a_2 \ne 0, \quad a_1 + a_2 + a_3 > 0 \text{ and } 0 \le a_3 - a_5, \quad (3.30)$$

$$E(x, y) \le a_4 d(x, y) + a_5 d(x, T^2 x), \quad (3.31)$$

where

$$E(x, y) := a_1 d(Tx, Ty) + a_2\big[d(x, Tx) + d(y, Ty)\big] + a_3[d(y, Tx) + d(x, Ty)],$$

hold for each $x, y \in X$. Then, for each $x \in X$, the sequence $\{T^n x\}$ converges to x^*. In addition, the limit x^* is the required fixed point of the selfmapping T.

3.7   Fixed Point Results via Admissible Mappings

In 2012, Samet et al. [254] proposed to use an admissible mapping to combine and to unify a vast number of the existing results on the topic in the literature. It has been studied very densely since then, see, e.g., [15, 17, 19, 21, 22, 32, 41, 43, 44, 93, 127, 145, 148, 174, 190].

First, we recall the notion of α−admissible mapping.

Definition 3.4 Let $\alpha : X \times X \to [0, \infty)$. A self-mapping $T : X \to X$ is called α−admissible if the condition

$$\alpha(x, y) \ge 1 \implies \alpha(Tx, Ty) \ge 1,$$

is satisfied for each $x, y \in X$.

Example [cf. [254]]

Let $X=(0, +\infty)$. Define $T : (0, +\infty) \rightarrow (0, +\infty)$ and $\alpha : (0, +\infty) \times (0, +\infty) \rightarrow [0, \infty)$ by

(1) Set $T(x) = \ln(x + 1)$, for every $x \in X$ and $\alpha(x, y) = \begin{cases} 0, \text{ if } x < y, \\ 2, \text{ if } x \geq y. \end{cases}$

So, T forms an α-admissible mapping.

(2) Set $T(x) = \sqrt[4]{x}$, for every $x \in X$ and $\alpha(x, y) = \begin{cases} 0, & \text{ if } x < y, \\ e^{x-y}, & \text{ if } x \geq y. \end{cases}$

Thus, T forms an α-admissible mapping.

Definition 3.5 (*See* [254]) Let X be any nonempty set, equipped with a metric d, and T be a selfmapping on X. If there are $\psi \in \Psi$ and $\alpha : X \times X \rightarrow [0, \infty)$ so that

$$\alpha(x, y)d(Tx, Ty) \leq \psi(d(x, y)), \text{ for each } x, y \in X,$$

then T is called an $\alpha - \psi$ contractive mapping.

It is evident that, each contraction, is an $\alpha - \psi$ contractive mapping by employing $\psi(t) = kt$, $k \in (0, 1)$, and $\alpha(x, y) = 1$ for each $x, y \in X$. For more examples and discussion of such type mappings, we refer to [174, 254].

Definition 3.6 A non-empty set X is strong-regular if for each recursive sequence $\{x_n\}$ is in X provides that $\alpha(x_n, x_{n+1}) \geq 1$ for each n and $x_n \rightarrow x \in X$ as $n \rightarrow \infty$, then there is a subsequence $\{x_{k(n)}\}$ of $\{x_n\}$ so that $\alpha(x_{k(n)}, x) \geq 1$ for each n.

For the sake of completeness, the main results in [254] is stated below:

Theorem 3.43 (See [254]) *We presume that the self-mapping T on a complete metric space (X, d) is an α-admissible and an $\alpha - \psi$ contractive mapping. Assume also that there is an initial point $x_0 \in X$ so that $\alpha(x_0, Tx_0) \geq 1$. If either, X is strong-regular, or, T is continuous, then T possesses a fixed point.*

Theorem 3.44 *In addition to the statements of Theorem 3.43, suppose the following condition is fulfilled: For any pair $(x, y) \in X \times X$, there is $z \in X$ so that $\alpha(x, z) \geq 1$ and $\alpha(y, z) \geq 1$. Then, the uniqueness of the existing fixed point of Theorem 3.43 is concluded.*

This result was extended in [174] as defining a generalized $\alpha - \psi$ contractive mappings:

Definition 3.7 For a metric space (X, d), a selfmapping $T : X \to X$ is called a generalized $\alpha - \psi$ contractive mapping if there exist $\psi \in \Psi$ and $\alpha : X \times X \to [0, \infty)$ so that

$$\alpha(x, y)d(Tx, Ty) \leq \psi(M(x, y)), \tag{3.32}$$

for each $x, y \in X$, where

$$M(x, y) = \max \left\{ d(x, y), \frac{d(x, Ty) + d(y, Tx)}{2}, \frac{d(x, Tx) + d(y, Ty)}{2} \right\}. \tag{3.33}$$

The first result of [174] is the following.

Theorem 3.45 *We presume that the self-mapping T on a complete metric space (X, d) forms an α-admissible and a generalized $\alpha - \psi$ contractive mapping. Assume also that there is an initial point $x_0 \in X$ so that $\alpha(x_0, Tx_0) \geq 1$. If T is continuous, then T possesses a fixed point.*

Proof Since there is an initial point $x_0 \in X$ so that $\alpha(x_0, Tx_0) \geq 1$, we can built a recursive sequence $\{x_n\}$ in X by using the rule $x_{n+1} = Tx_n$ for each $n \geq 0$. Note that in case of $x_{n_0} = x_{n_0+1}$ for some n_0, then the proof is terminated immediately since $x^* = x_{n_0}$ forms a fixed point of T. Based on this observation, for avoiding this simple case, we presume, for each $n \in \mathbb{N}_0$, that $x_n \neq x_{n+1}$. On account the fact that T is α-admissible, we find

$$\alpha(x_0, x_1) = \alpha(x_0, Tx_0) \geq 1 \text{ implies that } \alpha(Tx_0, Tx_1) = \alpha(x_1, x_2) \geq 1.$$

Recursively, we obtain

$$\alpha(x_n, x_{n+1}) \geq 1, \text{ for each } n \in \mathbb{N}_0. \tag{3.34}$$

Taking the expressions (3.32) and (3.58) into account, we deduce that

$$d(Tx_n, Tx_{n-1}) = d(x_{n+1}, x_n) \leq \alpha(x_n, x_{n-1})d(Tx_n, Tx_{n-1}) \leq \psi(M(x_n, x_{n-1})), \tag{3.35}$$

for each $n \in \mathbb{N}$. Observe, also, that

$$M(x_n, x_{n-1}) = \max \left\{ d(x_n, x_{n-1}), \frac{d(x_n, Tx_{n-1}) + d(x_{n-1}, Tx_n)}{2}, \frac{d(x_n, Tx_n) + d(x_{n-1}, Tx_{n-1})}{2} \right\}$$

$$= \max \left\{ d(x_n, x_{n-1}), \frac{d(x_{n-1}, x_{n+1})}{2}, \frac{d(x_n, x_{n+1}) + d(x_{n-1}, x_n)}{2} \right\}$$

$$\leq \max \left\{ d(x_n, x_{n-1}), \frac{d(x_n, x_{n+1}) + d(x_{n-1}, x_n)}{2} \right\}$$

$$\leq \max\{d(x_n, x_{n-1}), d(x_n, x_{n+1})\}.$$

Employing (3.59) and regarding the fact that the function ψ is increasing, we conclude

$$d(x_{n+1}, x_n) \leq \psi \left(\max\{d(x_n, x_{n-1}), d(x_n, x_{n+1})\} \right), \tag{3.36}$$

for any $n \in \mathbb{N}$. If $n_0 \in \mathbb{N}$ the inequality $d(x_{n_0}, x_{n_0-1}) \leq d(x_{n_0}, x_{n_0+1})$ holds, then from (3.36), we find that

$$d(x_{n_0+1}, x_{n_0}) \leq \psi(d(x_{n_0}, x_{n_0+1})) < d(x_{n_0}, x_{n_0+1}),$$

a contradiction. Consequently, for each $n \in \mathbb{N}$, we get

$$d(x_n, x_{n-1}) = \max\{d(x_n, x_{n-1}), d(x_n, x_{n+1})\}. \tag{3.37}$$

Combining (3.36) and (3.37), we deduce that

$$d(x_{n+1}, x_n) \leq \psi(d(x_n, x_{n-1})), \tag{3.38}$$

for each $n \in \mathbb{N}$. Recursively, we have

$$d(x_{n+1}, x_n) \leq \psi^n(d(x_1, x_0)), \text{ for each } n \in \mathbb{N}. \tag{3.39}$$

Due to the triangle inequality and the estimation (3.39), for each $k \in \mathbb{N}$, we find

$$\begin{aligned}
d(x_n, x_{n+k}) &\leq d(x_n, x_{n+1}) + \cdots + d(x_{n+k-1}, x_{n+k}) \\
&\leq \sum_{p=n}^{n+k-1} \psi^n(d(x_1, x_0)) \\
&\leq \sum_{p=n}^{+\infty} \psi^n(d(x_1, x_0)) \to 0 \text{ as } n \to \infty.
\end{aligned}$$

Consequently, the sequence $\{x_n\}$ is Cauchy in (X, d). Regarding the fact that (X, d) is complete, there is $x^* \in X$ so that

$$\lim_{n \to \infty} d(x_n, x^*) = 0. \tag{3.40}$$

Keeping the continuity assumption of T in mind, the expression (3.73) yields that

$$\lim_{n \to \infty} d(x_{n+1}, Tx^*) = \lim_{n \to \infty} d(Tx_n, Tx^*) = 0. \tag{3.41}$$

The uniqueness of the limit, together with (3.73) and (3.74) imply that x^* is a fixed point of T. $\qquad\square$

Definition 3.8 A non-empty set X is regular if for each recursive sequence $\{x_n\}$ is in X provides that $\alpha(x_n, x_{n+1}) \geq 1$ for each n and $x_n \to x \in X$ as $n \to \infty$, then there is a subsequence $\{x_{n(k)}\}$ of $\{x_n\}$ so that $\alpha(x_{n(k)}, x) \geq 1$ for each k.

In the next result, the continuity assumption of Theorem 3.46 is replaced with a weakened condition.

Theorem 3.46 *We presume that the self-mapping T on a complete metric space (X, d) forms an α-admissible and a generalized $\alpha - \psi$ contractive mapping. Assume also that there is an initial point $x_0 \in X$ so that $\alpha(x_0, Tx_0) \geq 1$. If X is regular, then T possesses a fixed point.*

Proof Verbatim of the proof of Theorem 3.45 yields that there is a recursive sequence $\{x_n\}$ converges for some $x^* \in X$, where $x_{n+1} = Tx_n$ for each $n \in \mathbb{N}_0$, . Since X is regular, by (3.58), there is a subsequence $\{x_{n(k)}\}$ of $\{x_n\}$ so that $\alpha(x_{n(k)}, x^*) \geq 1$ for each k. Applying (3.32), for each k, we find

$$d(x_{n(k)+1}, Tx^*) = d(Tx_{n(k)}, Tx^*) \leq \alpha(x_{n(k)}, x^*)d(Tx_{n(k)}, Tx^*)$$
$$\leq \psi(M(x_{n(k)}, x^*)). \tag{3.42}$$

Further, we observe that

$$M(x_{n(k)}, x^*) = \max\left\{ d(x_{n(k)}, x^*), \frac{d(x_{n(k)}, Tx^*) + d(x^*, x_{n(k)+1})}{2} \frac{d(x_{n(k)}, x_{n(k)+1}) + d(x^*, Tx^*)}{2} \right\}.$$

Setting $k \to \infty$ in the equality above, we deduce that

$$\lim_{k \to \infty} M(x_{n(k)}, x^*) = \frac{d(x^*, Tx^*)}{2}. \tag{3.43}$$

Assume that $d(x^*, Tx^*) > 0$. From (3.43), for k large enough, we have $M(x_{n(k)}, x^*) > 0$ that yields $\psi(M(x_{n(k)}, x^*)) < M(x_{n(k)}, x^*)$. Therefore, by (3.76), we get

$$d(x_{n(k)+1}, Tu) < M(x_{n(k)}, x^*).$$

Setting $k \to \infty$ in the inequality above, by employing (3.43), we find

$$d(x^*, Tx^*) \leq \frac{d(x^*, Tx^*)}{2},$$

a contradiction. Accordingly, we find $d(x^*, Tx^*) = 0$, that is, $x^* = Tx^*$. □

It is worth mentioning that Theorem 3.45 (resp. Theorem 3.46) guarantees only the existence of a fixed point. For the uniqueness, we need an additional condition. For this purpose, we propose the following hypothesis:

(H) For every $x, y \in \text{Fix}(T)$, there is $z \in X$ satisfying $\alpha(x, z) \geq 1$ and $\alpha(y, z) \geq 1$.

Here, $\text{Fix}(T)$ denotes the set of fixed points of T.

Theorem 3.47 *In addition to the statements of Theorem 3.45 (resp. Theorem 3.46), if the condition (H) is fulfilled, then we conclude that fixed point x^* of T is unique.*

Proof Suppose, on the contrary, that $y^* = Ty^*$ and $v \neq x^*$. From (H), there is $\omega \in X$ so that

$$\alpha(x^*, \omega) \geq 1 \text{ and } \alpha(v, \omega) \geq 1. \tag{3.44}$$

Keeping on mind that T is $\alpha-$admissible, (3.44) yields

$$\alpha(x^*, T^n\omega) \geq 1 \text{ and } \alpha(y^*, T^n\omega) \geq 1, \text{ for each } n. \tag{3.45}$$

We construct the sequence $\{\omega_n\}$ in X by $\omega_{n+1} = T\omega_n$ for each $n \geq 0$ and $\omega_0 = z$. From (3.45), for each n, we have

$$d(x^*, \omega_{n+1}) = d(Tx^*, T\omega_n) \leq \alpha(x^*, \omega_n)d(Tx^*, T\omega_n) \leq \psi(M(x^*, \omega_n)). \tag{3.46}$$

In addition, we observe that

$$
\begin{aligned}
M(x^*, \omega_n) &= \max\left\{d(x^*, \omega_n), \frac{d(\omega_n, \omega_{n+1})}{2}, \frac{d(x^*, \omega_{n+1}) + d(\omega_n, x^*)}{2}\right\} \\
&\leq \max\left\{d(x^*, \omega_n), \frac{d(\omega_n, x^*) + d(x^*, \omega_{n+1})}{2}\right\} \\
&\leq \max\{d(x^*, \omega_n), d(x^*, \omega_{n+1})\}.
\end{aligned}
$$

On account of the monotone property of ψ, together with the inequality above and (3.46), we find

$$d(x^*, \omega_{n+1}) \leq \psi\left(\max\{d(x^*, \omega_n), d(x^*, \omega_{n+1})\}\right), \tag{3.47}$$

for each n. Without restriction to the generality, we presume that $d(x^*, \omega_n) > 0$ for each n. If $\max\{d(x^*, \omega_n), d(x^*, \omega_{n+1})\} = d(x^*, \omega_{n+1})$, we obtain from (3.47) that

$$d(x^*, \omega_{n+1}) \leq \psi(d(x^*, \omega_{n+1})) < d(x^*, \omega_{n+1}),$$

a contradiction. Consequently, we find $\max\{d(x^*, \omega_n), d(x^*, \omega_{n+1})\} = d(x^*, \omega_n)$, and

$$d(x^*, \omega_{n+1}) \leq \psi(d(x^*, \omega_n)),$$

for each n. Hence, we obtain

$$d(x^*, \omega_n) \leq \psi^n(d(x^*, \omega_0)), \text{ for each } n \geq 1. \tag{3.48}$$

Setting $n \to \infty$ in (3.48), we find

$$\lim_{n \to \infty} d(\omega_n, x^*) = 0. \tag{3.49}$$

In an analogous way, we derive that

$$\lim_{n \to \infty} d(\omega_n, y^*) = 0. \tag{3.50}$$

Combining (3.49) and (3.50) yields that $x^* = y^*$. Accordingly, we indicated that x^* is the sole fixed point of T. □

Remark 3.1 From the statements and the proofs of Theorem 3.43, Theorems 3.44–3.46, we observe that the function $M(x, y)$ in (3.33) can be replaced by several other expressions considered in Sect. 3.2. In particular

$$\alpha(x, y)d(Tx, Ty) \leq \psi(M(x, y)),$$

can be replaced by

(O1.) inspired by Theorem 3.17; $\alpha(x, y)d(Tx, Ty) \leq \beta(d(x, y))d(x, y)$, or,

(O2.) inspired by Theorem 3.17; $\alpha(x, y)d(Tx, Ty) \leq \beta(M(x, y))M(x, y)$, or,

(O3.) inspired by $(N1)$ of Theorem 3.18;

$$\alpha(x, y)d(Tx, Ty) \leq k_1 \frac{[1 + d(x, Tx)]d(y, Ty)}{1 + d(x, y)} + k_2 d(x, y),$$

where, $k_1 + k_2 \in [0, 1)$, and k_1, k_2 are nonnegative reals, or,

(O3.) inspired by $(N1)$ of Theorem 3.18;

$$\alpha(x, y)d(Tx, Ty) \leq k \max\{\frac{[1 + d(x, Tx)]d(y, Ty)}{1 + d(x, y)}, d(x, y)\},$$

for $k \in [0, 1)$, or,

(O4.) inspired by $(N3)$ Theorem 3.18; $\alpha(x, y)d(Tx, Ty) \leq d(x, y) - \varphi(d(x, y))$, or,

(O5.) inspired by Suzuki [268–270];

$$\frac{1}{2}d(x, Tx) \le d(x, y)\alpha(x, y)d(Tx, Ty) \le \psi(M(x, y))$$

or many other expression and their combinations.

In the following subsection we shall demonstrate the immediate consequences of Theorem 3.47. It is evident that possible new extensions of Theorem 3.47 (as it discussed in Remark 3.1) will have such consequences, too.

3.7.1 Standard Fixed Point Theorems

In this subsection, we aim to list some of the immediate consequences of Theorem 3.47 in the setting of standard metric spaces. Indeed, the consequences of this subsection are mostly well-known existing results in the metric fixed point theory literature. The first corollary is obtained by letting $\alpha(x, y) = 1$, for each $x, y \in X$, in the statement of Theorem 3.47.

Corollary 3.6 *We presume that the self-mapping T on a complete metric space (X, d). If there is a (c)-comparison function $\psi \in \Psi$ so that*

$$d(Tx, Ty) \le \psi(M(x, y)),$$

for each $x, y \in X$, then the selfmapping T possesses a unique fixed point.

The following result is derived from Corollary 3.6 by keeping the monotonicity of the auxiliary function in mind.

Corollary 3.7 *If there is a (c)-comparison function $\psi \in \Psi$ so that*

$$d(Tx, Ty) \le \psi\left(\max\left\{\frac{d(x, Tx) + d(y, Ty)}{2}, \frac{d(x, Ty) + d(y, Tx)}{2}\right\}\right),$$

for each $x, y \in X$, then the selfmapping T possesses a unique fixed point.

Sketch the Proof Owing to the fact (c)-comparison function ψ is nondecreasing together with the following inequality,

$$\max\left\{\frac{d(x, Tx) + d(y, Ty)}{2}, \frac{d(x, Ty) + d(y, Tx)}{2}\right\} \le M(x, y),$$

for each $x, y \in X$, we derive the result.

As in the previous consequence, by help of the monotonicity of ψ, we get the following result, from Corollary 3.6, which is inspired by Reich's type fixed point theorem:

Corollary 3.8 *If there is a (c)-comparison function $\psi \in \Psi$ so that*

$$d(Tx, Ty) \le \psi(\max\left\{d(x, y), \frac{d(x, Tx) + d(y, Ty)}{2}\right\}),$$

for each $x, y \in X$, then the selfmapping T possesses a unique fixed point.

Sketch the Proof Due to the fact (c)-comparison function ψ is nondecreasing together with the following inequality,

$$\max\left\{d(x, y), \frac{d(x, Tx) + d(y, Ty)}{2}\right\} \le M(x, y),$$

for each $x, y \in X$, we derive the result.

Corollary 3.9 *If there is a (c)-comparison function $\psi \in \Psi$ so that*

$$d(Tx, Ty) \le \psi(\max\left\{d(x, y), \frac{d(x, Ty) + d(y, Tx)}{2}\right\}),$$

for each $x, y \in X$, then the selfmapping T possesses a unique fixed point.

Sketch the proof. Combining the fact (c)-comparison function ψ is nondecreasing together with the following inequality,

$$\max\left\{d(x, y), \frac{d(x, Ty) + d(y, Tx)}{2}\right\} \le M(x, y),$$

for each $x, y \in X$, we derive the result.

Keeping the Kannan's fixed point theorem in mind, Corollary 3.6 implies the following result:

Corollary 3.10 *If there is a (c)-comparison function $\psi \in \Psi$ so that*

$$d(Tx, Ty) \le \psi(\frac{d(x, Tx) + d(y, Ty)}{2}),$$

for each $x, y \in X$, then the selfmapping T possesses a unique fixed point.

Sketch the Proof Combining the fact (c)-comparison function ψ is nondecreasing together with the following inequality,

$$\frac{d(x, Tx) + d(y, Ty)}{2} \leq M(x, y),$$

for each $x, y \in X$, we derive the result.

Corollary 3.6 yields also the following result that is inspired from Chaterjea's fixed point theorem.

Corollary 3.11 *If there is a (c)-comparison function $\psi \in \Psi$ so that*

$$d(Tx, Ty) \leq \psi(\frac{d(x, Ty) + d(y, Tx)}{2}),$$

for each $x, y \in X$, then the selfmapping T possesses a unique fixed point.

Sketch the Proof Combining the fact (c)-comparison function ψ is nondecreasing together with the following inequality,

$$\frac{d(x, Ty) + d(y, Tx)}{2} \leq M(x, y),$$

for each $x, y \in X$, we derive the result.

Corollary 3.6 yields the following corollary which is known as Berinde's fixed point theorem [61].

Corollary 3.12 (see Berinde [61]) *If there is a (c)-comparison function $\psi \in \Psi$ so that*

$$d(Tx, Ty) \leq \psi(d(x, y)),$$

for each $x, y \in X$, then the selfmapping T possesses a unique fixed point.

Sketch the proof. Combining the fact (c)-comparison function ψ is nondecreasing together with the following inequality,

$$d(x, y) \leq M(x, y),$$

for each $x, y \in X$, we derive the result.

Corollary 3.6 implies also the following corollary which is known as Ćirić's fixed point theorem [100].

Corollary 3.13 (see Ćirić [100]) *We presume that the self-mapping T on a complete metric space (X, d). If there is a real number $\lambda \in (0, 1)$ so that*

$$d(Tx, Ty) \leq \lambda \max \left\{ d(x, y), \frac{d(x, Tx) + d(y, Ty)}{2}, \frac{d(x, Ty) + d(y, Tx)}{2} \right\},$$

for each $x, y \in X$, then the selfmapping T possesses a unique fixed point.

Sketch of the proof. It is sufficient to take $\psi(t) = \lambda t$, where $\lambda \in (0, 1)$. Evidently, $\psi(t)$ forms a (c)-comparison function, hence, Corollary 3.6 provides the desired result.

Corollary 3.13 yields the following corollary which is known as Hardy and Rogers's fixed point theorem [128]

Corollary 3.14 (see Hardy and Rogers [128]) *We presume that the self-mapping T on a complete metric space (X, d). If there are nonnegative real numbers $A, B, C \geq 0$ with $(A + 2B + 2C) \in (0, 1)$ so that*

$$d(Tx, Ty) \leq Ad(x, y) + B[d(x, Tx) + d(y, Ty)] + C[d(x, Ty) + d(y, Tx)],$$

for each $x, y \in X$, then the selfmapping T possesses a unique fixed point.

Sketch of the Proof For the sake of simplicity, let us rename the expressions in Corollary 3.14, as follows:

$$C(x, y) := \lambda \max \left\{ d(x, y), \frac{1}{2}[d(x, Tx) + d(y, Ty)], \frac{1}{2}[d(x, Ty) + d(y, Tx)] \right\},$$

and $HR(x, y) := Ad(x, y) + B[d(x, Tx) + d(y, Ty)] + C[d(x, Ty) + d(y, Tx)]$. It is evident that $HR(x, y) \leq C(x, y)$ for each $x, y \in X$. Thus, Corollary 3.13 yields the desired result.

The following is the most interesting consequences of Corollary 3.12 (implicitly, Theorem 3.47)

Corollary 3.15 (Banach Contraction Principle [56]) *We presume that the self-mapping T on a complete metric space (X, d). If there is a real number $\lambda \in (0, 1)$ so that*

$$d(Tx, Ty) \leq \lambda d(x, y),$$

for each $x, y \in X$, then the selfmapping T possesses a unique fixed point.

Sketch of the proof. It is sufficient to take $\psi(t) = \lambda t$, where $\lambda \in (0, 1)$. It is evident that $\psi(t)$ forms a (c)-comparison function. Accordingly, Corollary 3.12 concludes the desired result.

Another renowned consequence of Theorem 3.47 (indeed, Corollary 3.13) is the well-known Kannan's fixed point theorem [183].

Corollary 3.16 (see Kannan [183]) *We presume that the self-mapping T on a complete metric space (X, d). If there is a real number $\lambda \in (0, 1)$ so that*

$$d(Tx, Ty) \le \frac{\lambda}{2}[d(x, Tx) + d(y, Ty)],$$

for each $x, y \in X$, then the selfmapping T possesses a unique fixed point.

Sketch of the Proof It is derived from Corollary 3.13 since

$$\frac{\lambda}{2}[d(x, Tx) + d(y, Ty)] \le C(x, y),$$

for each $x, y \in X$.

The famous Chatterjea's fixed point theorem [89] is also an immediate consequence of Theorem 3.47 (indeed, Corollary 3.13).

Corollary 3.17 (see Chatterjea [89]) *We presume that the self-mapping T on a complete metric space (X, d). If there is a real number $\lambda \in (0, 1)$ so that*

$$d(Tx, Ty) \le \frac{\lambda}{2}[d(x, Ty) + d(y, Tx)],$$

for each $x, y \in X$, then the selfmapping T possesses a unique fixed point.

Sketch of the Proof It is derived from Corollary 3.13 since

$$\frac{\lambda}{2}[d(x, Ty) + d(y, Tx)] \le C(x, y),$$

for each $x, y \in X$.

3.7.2 Fixed Point Theorems on Metric Spaces Endowed with a Partial Order

In this subsection, we discuss the consequences of Theorem 3.47 in the framework of partially ordered metric spaces. We first recall some basic notions that will be used in this subsection.

Definition 3.9 Let X be a nonempty set equipped with a partial order $\preceq$. A selfmapping T over X is said to be nondecreasing with respect to the partial order $\preceq$ whenever

$$x, y \in X, \ x \preceq y \Longrightarrow Tx \preceq Ty.$$

In addition, we say that a sequence $\{x_n\} \subset X$ is nondecreasing with respect to $\preceq$ whenever $x_n \preceq x_{n+1}$ for each n.

Definition 3.10 Let X be a nonempty set equipped with a partial order $\preceq$ and metric d. A partially ordered metric space $(X, \preceq, d)$ is called regular whenever for every nondecreasing sequence $\{x_n\} \subset X$ so that $x_n \to x \in X$ as $n \to \infty$, there is a subsequence $\{x_{n(k)}\}$ of $\{x_n\}$ so that $x_{n(k)} \preceq x$ for each k.

We have the following result.

Corollary 3.18 *Let X be a nonempty set equipped with a partial order $\preceq$ and metric d. Suppose the corresponding metric space (X, d) is complete. Further, $T : X \to X$ is nondecreasing mapping with respect to $\preceq$. We also presume that*

(i)   *there is $x_0 \in X$ so that $x_0 \preceq T x_0$;*
(ii) *there is a (c)-comparison function $\psi \in \Psi$ so that*

$$d(Tx, Ty) \leq \psi(M(x, y)),$$

    for every $x, y \in X$ with $x \succeq y$;
(iii) *either $(X, \preceq, d)$ is regular or T is continuous.*

Then, the selfmapping T possesses a fixed point. Furthermore, this fixed point is unique if for each $x, y \in X$ there is $z \in X$ so that $y \preceq z$ and $x \preceq z$.

Sketch of the Proof It is sufficient to define the auxiliary mapping α in connection of the partial relation $\preceq$ to conclude that Corollary 3.18 is an immediate consequences of Theorem 3.47. Set $\alpha : X \times X \to [0, \infty)$ by

$$\alpha(x, y) = \begin{cases} 1 \text{ if } x \preceq y \text{ or } x \succeq y, \\ 0 \text{ otherwise.} \end{cases}$$

It is evident that all axioms of Theorem 3.47 are fulfilled according to the α defined above.

    Indeed, by defining α as above, we state all corollaries in the previous subsection in the context of metric spaces endowed with a partial order.

Corollary 3.19 *Let X be a nonempty set equipped with a partial order $\preceq$ and metric d. Suppose the corresponding metric space (X, d) is complete. Further, $T : X \to X$ is nondecreasing mapping with respect to $\preceq$. We also presume that*

(i) *there is $x_0 \in X$ so that $x_0 \preceq T x_0$;*
(ii) *there is a (c)-comparison function $\psi \in \Psi$ so that*

$$d(Tx, Ty) \leq \psi \left(\max \left\{ \frac{d(x, Tx) + d(y, Ty)}{2}, \frac{d(x, Ty) + d(y, Tx)}{2} \right\} \right);$$

(iii) either $(X, \preceq, d)$ is regular or T is continuous.

Then, the selfmapping T possesses a fixed point. Furthermore, this fixed point is unique if for each $x, y \in X$ there is $z \in X$ so that $y \preceq z$ and $x \preceq z$.

Corollary 3.20 *Let X be a nonempty set equipped with a partial order $\preceq$ and metric d. Suppose the corresponding metric space (X, d) is complete. Further, $T : X \rightarrow X$ is nondecreasing mapping with respect to $\preceq$. We also presume that*

(i) *there is $x_0 \in X$ so that $x_0 \preceq Tx_0$;*
(ii) *there is a (c)-comparison function $\psi \in \Psi$ so that*

$$d(Tx, Ty) \leq \psi \left(\max \left\{ d(x, y), \frac{d(x, Ty) + d(y, Tx)}{2} \right\} \right);$$

(iii) *either $(X, \preceq, d)$ is regular or T is continuous.*

Then, the selfmapping T possesses a fixed point. Furthermore, this fixed point is unique if for each $x, y \in X$ there is $z \in X$ so that $y \preceq z$ and $x \preceq z$.

Corollary 3.21 *Let X be a nonempty set equipped with a partial order $\preceq$ and metric d. Suppose the corresponding metric space (X, d) is complete. Further, $T : X \rightarrow X$ is nondecreasing mapping with respect to $\preceq$. We also presume that*

(i) *there is $x_0 \in X$ so that $x_0 \preceq Tx_0$;*
(ii) *there is a (c)-comparison function $\psi \in \Psi$ so that*

$$d(Tx, Ty) \leq \psi \left(\max \left\{ d(x, y), \frac{d(x, Ty) + d(y, Tx)}{2} \right\} \right);$$

(iii) *either $(X, \preceq, d)$ is regular or T is continuous.*

Then, the selfmapping T possesses a fixed point. Furthermore, this fixed point is unique if for each $x, y \in X$ there is $z \in X$ so that $y \preceq z$ and $x \preceq z$.

Corollary 3.22 *Let X be a nonempty set equipped with a partial order $\preceq$ and metric d. Suppose the corresponding metric space (X, d) is complete. Further, $T : X \rightarrow X$ is nondecreasing mapping with respect to $\preceq$. We also presume that*

(i) *there is $x_0 \in X$ so that $x_0 \preceq Tx_0$;*

(ii) there is a (c)-comparison function $\psi \in \Psi$ so that

$$d(Tx, Ty) \leq \psi(\frac{d(x, Tx) + d(y, Ty)}{2});$$

(iii) either $(X, \preceq, d)$ is regular or T is continuous.

Then, the selfmapping T possesses a fixed point. Furthermore, this fixed point is unique if for each $x, y \in X$ there is $z \in X$ so that $y \preceq z$ and $x \preceq z$.

Corollary 3.23 *Let X be a nonempty set equipped with a partial order $\preceq$ and metric d. Suppose the corresponding metric space (X, d) is complete. Further, $T : X \to X$ is nondecreasing mapping with respect to $\preceq$. We also presume that*

(i)   there is $x_0 \in X$ so that $x_0 \preceq Tx_0$;
(ii) there is a (c)-comparison function $\psi \in \Psi$ so that

$$d(Tx, Ty) \leq \psi(\frac{d(x, Ty) + d(y, Tx)}{2});$$

(iii) either $(X, \preceq, d)$ is regular or T is continuous.

Then, the selfmapping T possesses a fixed point. Furthermore, this fixed point is unique if for each $x, y \in X$ there is $z \in X$ so that $y \preceq z$ and $x \preceq z$.

Corollary 3.24 *Let X be a nonempty set equipped with a partial order $\preceq$ and metric d. Suppose the corresponding metric space (X, d) is complete. Further, $T : X \to X$ is nondecreasing mapping with respect to $\preceq$. We also presume that*

(i)   there is $x_0 \in X$ so that $x_0 \preceq Tx_0$;
(ii) there is a (c)-comparison function $\psi \in \Psi$ so that

$$d(Tx, Ty) \leq \psi(d(x, y));$$

(iii) either $(X, \preceq, d)$ is regular or T is continuous.

Then, the selfmapping T possesses a fixed point. Furthermore, this fixed point is unique if for each $x, y \in X$ there is $z \in X$ so that $y \preceq z$ and $x \preceq z$.

Corollary 3.25 *Let X be a nonempty set equipped with a partial order $\preceq$ and metric d. Suppose the corresponding metric space (X, d) is complete. Further, $T : X \to X$ is nondecreasing mapping with respect to $\preceq$. We also presume that*

(i) *there is $x_0 \in X$ so that $x_0 \preceq Tx_0$;*
(ii) *there is a real number $\lambda \in (0, 1)$ so that*

$$d(Tx, Ty) \le \lambda \max \left\{ d(x, y), \frac{d(x, Tx) + d(y, Ty)}{2}, \frac{d(x, Ty) + d(y, Tx)}{2} \right\};$$

(iii) *either $(X, \preceq, d)$ is regular or T is continuous.*

Then, the selfmapping T possesses a fixed point. Furthermore, this fixed point is unique if for each $x, y \in X$ there is $z \in X$ so that $y \preceq z$ and $x \preceq z$.

Corollary 3.26 *Let X be a nonempty set equipped with a partial order $\preceq$ and metric d. Suppose the corresponding metric space (X, d) is complete. Further, $T : X \to X$ is nondecreasing mapping with respect to $\preceq$. We also presume that*

(i) *there is $x_0 \in X$ so that $x_0 \preceq Tx_0$;*
(ii) *there exist real numbers $A, B, C \ge 0$ with $(A + 2B + 2C) \in (0, 1)$ so that*

$$d(Tx, Ty) \le Ad(x, y) + B[d(x, Tx) + d(y, Ty)] + C[d(x, Ty) + d(y, Tx)];$$

(iii) *either $(X, \preceq, d)$ is regular or T is continuous.*

Then, the selfmapping T possesses a fixed point. Furthermore, this fixed point is unique if for each $x, y \in X$ there is $z \in X$ so that $y \preceq z$ and $x \preceq z$.

Corollary 3.27 (see Ran and Reurings [235], Nieto and López [221]) *Let X be a nonempty set equipped with a partial order $\preceq$ and metric d. Suppose the corresponding metric space (X, d) is complete. Further, $T : X \to X$ is nondecreasing mapping with respect to $\preceq$. We also presume that*

(i) *there is $x_0 \in X$ so that $x_0 \preceq Tx_0$;*
(ii) *there is a real number $\lambda \in (0, 1)$ so that*

$$d(Tx, Ty) \le \lambda d(x, y);$$

(iii) *either $(X, \preceq, d)$ is regular or T is continuous.*

Then, the selfmapping T possesses a fixed point. Furthermore, this fixed point is unique if for each $x, y \in X$ there is $z \in X$ so that $y \preceq z$ and $x \preceq z$.

Corollary 3.28 *Let X be a nonempty set equipped with a partial order $\preceq$ and metric d. Suppose the corresponding metric space (X, d) is complete. Further, $T : X \to X$ is nondecreasing mapping with respect to $\preceq$. We also presume that*

(i)   there is $x_0 \in X$ so that $x_0 \preceq Tx_0$;
(ii) there is a real number $\lambda \in (0, 1)$ so that

$$d(Tx, Ty) \leq \frac{\lambda}{2}[d(x, Tx) + d(y, Ty)];$$

(iii) either $(X, \preceq, d)$ is regular or T is continuous.

Then, the selfmapping T possesses a fixed point. Furthermore, this fixed point is unique if for each $x, y \in X$ there is $z \in X$ so that $y \preceq z$ and $x \preceq z$.

Corollary 3.29 *Let X be a nonempty set equipped with a partial order $\preceq$ and metric d. Suppose the corresponding metric space (X, d) is complete. Further, $T : X \to X$ is nondecreasing mapping with respect to $\preceq$. We also presume that*

(i)   there is $x_0 \in X$ so that $x_0 \preceq Tx_0$;
(ii) there is a real number $\lambda \in (0, 1)$ so that

$$d(Tx, Ty) \leq \frac{\lambda}{2}[d(x, Ty) + d(y, Tx)];$$

(iii) either $(X, \preceq, d)$ is regular or T is continuous.

Then, the selfmapping T possesses a fixed point. Furthermore, this fixed point is unique if for each $x, y \in X$ there is $z \in X$ so that $y \preceq z$ and $x \preceq z$.

3.7.3   Fixed Point Theorems for Cyclic Contractive Mappings

In fixed point theory, finding fixed point for the discontinuous mapping takes attraction attention of researcher. One of earlier results in the direction was given by Kannan [152]. A mapping T may have a fixed point despite to the discontinuity of T whenever one of the iteration of T, for example, T^2 is continuous. Another novel result on the existing a fixed point for discontinuous mapping was introduced by Kirk et al. [197]. Indeed, Kirk et al. [197] defined cyclic contraction via cyclic mapping. After the pioneer result of Kirk et al. [197], a significant number of the researches published interesting results in this direction, see, for instance, [7, 163, 170, 171, 224, 227, 253].

The goal of this subsection, is to indicate that Theorem 3.47 covers corresponding fixed point theorems in the setting of cyclic contractions. The first corollary, in this direction, is the following.

Corollary 3.30 *For a complete metric space (X, d), we presume that $\{A_i\}_{i=1}^{2}$ are nonempty closed subsets of and $T : A_1 \cup A_2 \to A_1 \cup A_2$ is a given mapping with $T(A_1) \subseteq A_2$ and*

$T(A_2) \subseteq A_1$. *If there is a function* $\psi \in \Psi$ *so that*

$$d(Tx, Ty) \leq \psi(M(x, y)), \text{ for each } (x, y) \in A_1 \times A_2,$$

then T possesses a unique fixed point $x^* \in A_1 \cap A_2$.

Sketch of the Proof Since A_1 and A_2 are closed subsets of the complete metric space (X, d), then (Y, d) is complete. Define the mapping $\alpha : Y \times Y \rightarrow [0, \infty)$ by

$$\alpha(x, y) = \begin{cases} 1 \text{ if } (x, y) \in (A_1 \times A_2) \cup (A_2 \times A_1), \\ 0 \text{ otherwise.} \end{cases}$$

The rest is straightforward confirmation of the conditions of Theorem 3.47.

Corollary 3.31 *For a complete metric space* (X, d), *we presume that* $\{A_i\}_{i=1}^{2}$ *are nonempty closed subsets of and* $T : A_1 \cup A_2 \rightarrow A_1 \cup A_2$ *is a given mapping with* $T(A_1) \subseteq A_2$ *and* $T(A_2) \subseteq A_1$. *If there is a function* $\psi \in \Psi$ *so that*

$$d(Tx, Ty) \leq \psi(\max\left\{\frac{d(x, Ty) + d(y, Tx)}{2}, \frac{d(x, Tx) + d(y, Ty)}{2}\right\}), \text{ for each } (x, y) \in A_1 \times A_2,$$

then T possesses a unique fixed point $x^* \in A_1 \cap A_2$.

Corollary 3.32 *For a complete metric space* (X, d), *we presume that* $\{A_i\}_{i=1}^{2}$ *are nonempty closed subsets of and* $T : A_1 \cup A_2 \rightarrow A_1 \cup A_2$ *is a given mapping with* $T(A_1) \subseteq A_2$ *and* $T(A_2) \subseteq A_1$. *If there is a function* $\psi \in \Psi$ *so that*

$$d(Tx, Ty) \leq \psi(\max\left\{d(x, y), \frac{d(x, Tx) + d(y, Ty)}{2}\right\}), \text{ for each } (x, y) \in A_1 \times A_2,$$

then T possesses a unique fixed point $x^* \in A_1 \cap A_2$.

Corollary 3.33 *For a complete metric space* (X, d), *we presume that* $\{A_i\}_{i=1}^{2}$ *are nonempty closed subsets of and* $T : A_1 \cup A_2 \rightarrow A_1 \cup A_2$ *is a given mapping with* $T(A_1) \subseteq A_2$ *and* $T(A_2) \subseteq A_1$. *If there is a function* $\psi \in \Psi$ *so that*

$$d(Tx, Ty) \leq \psi(\max\left\{d(x, y), \frac{d(x, Ty) + d(y, Tx)}{2}\right\}), \text{ for each } (x, y) \in A_1 \times A_2,$$

then T possesses a unique fixed point $x^* \in A_1 \cap A_2$.

Corollary 3.34 *For a complete metric space* (X, d), *we presume that* $\{A_i\}_{i=1}^2$ *are nonempty closed subsets of and* $T : A_1 \cup A_2 \rightarrow A_1 \cup A_2$ *is a given mapping with* $T(A_1) \subseteq A_2$ *and* $T(A_2) \subseteq A_1$. *If there is a function* $\psi \in \Psi$ *so that*

$$d(Tx, Ty) \leq \psi(\frac{d(x, Tx) + d(y, Ty)}{2}), \text{ for each } (x, y) \in A_1 \times A_2,$$

then T *possesses a unique fixed point* $x^* \in A_1 \cap A_2$.

Corollary 3.35 *For a complete metric space* (X, d), *we presume that* $\{A_i\}_{i=1}^2$ *are nonempty closed subsets of and* $T : A_1 \cup A_2 \rightarrow A_1 \cup A_2$ *is a given mapping with* $T(A_1) \subseteq A_2$ *and* $T(A_2) \subseteq A_1$. *If there is a function* $\psi \in \Psi$ *so that*

$$d(Tx, Ty) \leq \psi(\frac{d(x, Ty) + d(y, Tx)}{2}), \text{ for each } (x, y) \in A_1 \times A_2,$$

then T *possesses a unique fixed point* $x^* \in A_1 \cap A_2$.

Corollary 3.36 (see Pacurar and Rus [224]) *For a complete metric space* (X, d), *we presume that* $\{A_i\}_{i=1}^2$ *are nonempty closed subsets of and* $T : A_1 \cup A_2 \rightarrow A_1 \cup A_2$ *is a given mapping with* $T(A_1) \subseteq A_2$ *and* $T(A_2) \subseteq A_1$. *If there is a function* $\psi \in \Psi$ *so that*

$$d(Tx, Ty) \leq \psi(d(x, y)), \text{ for each } (x, y) \in A_1 \times A_2,$$

then T *possesses a unique fixed point* $x^* \in A_1 \cap A_2$.

Corollary 3.37 *For a complete metric space* (X, d), *we presume that* $\{A_i\}_{i=1}^2$ *are nonempty closed subsets of and* $T : A_1 \cup A_2 \rightarrow A_1 \cup A_2$ *is a given mapping with* $T(A_1) \subseteq A_2$ *and* $T(A_2) \subseteq A_1$. *If there is a real number* $\lambda \in (0, 1)$ *so that*

$$d(Tx, Ty) \leq \lambda \max \left\{ d(x, y), \frac{d(x, Tx) + d(y, Ty)}{2}, \frac{d(x, Ty) + d(y, Tx)}{2} \right\},$$

for each $(x, y) \in A_1 \times A_2$, *then* T *possesses a unique fixed point* $x^* \in A_1 \cap A_2$.

Corollary 3.38 *For a complete metric space* (X, d), *we presume that* $\{A_i\}_{i=1}^2$ *are nonempty closed subsets of and* $T : A_1 \cup A_2 \rightarrow A_1 \cup A_2$ *is a given mapping with* $T(A_1) \subseteq A_2$ *and* $T(A_2) \subseteq A_1$. *If there is a real numbers* $A, B, C \geq 0$ *with* $(A + 2B + 2C) \in (0, 1)$ *so that*

$$d(Tx, Ty) \leq Ad(x, y) + B[d(x, Tx) + d(y, Ty)] + C[d(x, Ty) + d(y, Tx)],$$

for each $(x, y) \in A_1 \times A_2$, *then* T *possesses a unique fixed point* $x^* \in A_1 \cap A_2$.

Corollary 3.39 (see Kirk et al. [197]) *For a complete metric space* (X, d), *we presume that* $\{A_i\}_{i=1}^2$ *are nonempty closed subsets of and* $T : A_1 \cup A_2 \to A_1 \cup A_2$ *is a given mapping with* $T(A_1) \subseteq A_2$ *and* $T(A_2) \subseteq A_1$. *If there is a real number* $\lambda \in (0, 1)$ *so that*

$$d(Tx, Ty) \leq \lambda d(x, y), \text{ for each } (x, y) \in A_1 \times A_2,$$

then T *possesses a unique fixed point* $x^* \in A_1 \cap A_2$.

Corollary 3.40 *For a complete metric space* (X, d), *we presume that* $\{A_i\}_{i=1}^2$ *are nonempty closed subsets of and* $T : A_1 \cup A_2 \to A_1 \cup A_2$ *is a given mapping with* $T(A_1) \subseteq A_2$ *and* $T(A_2) \subseteq A_1$. *If there is a real number* $\lambda \in (0, 1)$ *so that*

$$d(Tx, Ty) \leq \frac{\lambda}{2}[d(x, Tx) + d(y, Ty)] \text{ for each } (x, y) \in A_1 \times A_2,$$

then T *possesses a unique fixed point* $x^* \in A_1 \cap A_2$.

Corollary 3.41 *For a complete metric space* (X, d), *we presume that* $\{A_i\}_{i=1}^2$ *are nonempty closed subsets of and* $T : A_1 \cup A_2 \to A_1 \cup A_2$ *is a given mapping with* $T(A_1) \subseteq A_2$ *and* $T(A_2) \subseteq A_1$. *If there is a real number* $\lambda \in (0, 1)$ *so that*

$$d(Tx, Ty) \leq \frac{\lambda}{2}[d(x, Ty) + d(y, Tx)] \text{ for each } (x, y) \in A_1 \times A_2,$$

then T *possesses a unique fixed point* $x^* \in A_1 \cap A_2$.

Observation Remark. In this section, we observe that the usage of the admissible mapping leads to combining their trends in fixed point theory in a single expression. In other words, the same contraction condition yields a fixed point in three distinct abstract structures:

a standard compete metric space,

a compete metric space equipped by a partial order,

a compete metric space in which the operators are cyclic.

Note that further consequences can be added at the end of these sections, but we aim to discuss only the well-known results. Ergo, admissible mappings combine these three structures in a single statement with three immediate consequences.

3.8   Fixed Point Results via Simulation Functions

In this section, we shall consider the simulation functions and fixed point results that are obtained by these auxiliary functions. We, first, recall the definition of the simulation function.

Definition 3.11 We say that $\zeta : [0, \infty) \times [0, \infty) \to \mathbb{R}$ is *simulation function* if

(ζ_1) $\zeta(t, s) < s - t$ for each $s, t > 0$;
(ζ_2) if the positive sequences $\{t_n\}, \{s_n\}$ satisfies $\lim\limits_{n \to \infty} t_n = \lim\limits_{n \to \infty} s_n > 0$, then

$$\limsup_{n \to \infty} \zeta(t_n, s_n) < 0. \tag{3.51}$$

We underline that in the initial form of the definition, given in [194], there was a superfluous hypothesis, $\zeta(0, 0) = 0$. Later, this hypothesis is omitted, see e.g. Argoubi et al. [31].

The letter $\mathcal{Z}$ represent the class of all simulation functions $\zeta : [0, \infty) \times [0, \infty) \to \mathbb{R}$, that is, verifying (ζ_1) and (ζ_2).

Due to (ζ_1), we deduce

$$\zeta(t, t) < 0 \text{ for each } t > 0. \tag{3.52}$$

In what follows, we collect the fundamental examples of simulation functions.

Example [24, 194, 242]

Let $\mu_i : \mathbb{R}_0^+ \to \mathbb{R}_0^+$ be continuous functions so that $\mu_i(t) = 0$ if and only if, $t = 0$. Consider the following functions $\zeta_i : \mathbb{R}_0^+ \times \mathbb{R}_0^+ \to \mathbb{R}$, for $i = 1, 2, 3, 4, 5, 6$:

(i) $\zeta_1(t, s) = \mu_1(s) - \mu_2(t)$ for each $t, s \in [0, \infty)$, where $\mu_1, \mu_2 : \mathbb{R}_0^+ \to \mathbb{R}_0^+$ are two continuous functions so that $\mu_1(t) = \mu_2(t) = 0$ if and only if $t = 0$ and $\mu_1(t) < t \le \mu_2(t)$ for each $t > 0$.
(ii) $\zeta_2(t, s) = s - \dfrac{f(t, s)}{g(t, s)} t$ for each $t, s \in [0, \infty)$, where $T, g : [0, \infty)^2 \to (0, \infty)$ are two continuous functions with respect to each variable so that $T(t, s) > g(t, s)$ for each $t, s > 0$.
(iii) $\zeta_3(t, s) = s - \mu_3(s) - t$ for each $t, s \in [0, \infty)$.
(iv) $\zeta_4(t, s) = s \varphi(s) - t$ for each $s, t \in [0, \infty)$, where $\varphi : [0, \infty) \to [0, 1)$ is a function so that $\limsup\limits_{t \to r^+} \varphi(t) < 1$ for each $r > 0$.
(v) $\zeta_5(t, s) = \eta(s) - t$ for each $s, t \in [0, \infty)$, where $\eta : \mathbb{R}_0^+ \to \mathbb{R}_0^+$ is an upper semi-continuous mapping so that $\eta(t) < t$ for each $t > 0$ and $\eta(0) = 0$.

(vi) $\zeta_6(t, s) = s - \int_0^t \mu(u)du$ for each $s, t \in [0, \infty)$, where $\mu : [0, \infty) \to [0, \infty)$ is a
function so that $\int_0^\varepsilon \mu(u)du$ exists and $\int_0^\varepsilon \mu(u)du > \varepsilon$, for each $\varepsilon > 0$.

Evidently, $\zeta_1 - \zeta_6$ are simulation functions.

A mapping T, from a metric space (X, d) onto itself, is called $\mathcal{Z}$-contraction with respect
to ζ [194], if, for each $x, y \in X$,

$$\zeta(d(Tx, Ty), d(x, y)) \geq 0. \tag{3.53}$$

Due to (ζ_2), we conclude that

$$d(Tx, Ty) \neq d(x, y) \text{ for each distinct } x, y \in X. \tag{3.54}$$

In other words, whenever T is $\mathcal{Z}$-contraction, it cannot be an isometry. It is equivalent to
say, in a metric space, that the existence of fixed point for a $\mathcal{Z}$-contraction T necessarily
implies the uniqueness of the fixed point.

Theorem 3.48 *Every $\mathcal{Z}$-contraction on a complete metric space has a unique fixed point.
In fact, every Picard sequence converges to its unique fixed point.*

Definition 3.12 A self-mapping T, defined on a metric space (X, d), is called an α-
admissible $\mathcal{Z}$-contraction with respect to ζ, if there are two functions $\zeta \in \mathcal{Z}$ and $\alpha :
X \times X \to [0, \infty)$ that fulfill

$$\zeta(\alpha(x, y)d(Tx, Ty), d(x, y)) \geq 0 \qquad \text{for each } x, y \in X. \tag{3.55}$$

Incase of $\alpha(x, y) = 1$, the selfmapping T becomes $\mathcal{Z}$-contraction with respect to ζ.

Remark 3.2 For any α-admissible $\mathcal{Z}$-contraction T with respect to ζ, we have

$$\alpha(x, y)d(Tx, Ty) < d(x, y) \quad \text{for each } x, y \in X. \tag{3.56}$$

Suppose $x \neq y$ that is $d(x, y) > 0$. Notice that for the case $Tx = Ty$, we have the
desired result, trivially: $\alpha(x, y)d(Tx, Ty) = 0 < d(x, y)$. For the case, $Tx \neq Ty$, we have
$d(Tx, Ty) > 0$. It is evident that the inequality is fulfilled trivially if $\alpha(x, y) = 0$. For this
reason, we presume that $\alpha(x, y) > 0$ Employing (ζ_2) with (3.77), we deduce that

$$0 \leq \zeta(\alpha(x, y)d(Tx, Ty), d(x, y)) < d(x, y) - \alpha(x, y)d(Tx, Ty).$$

Thus, we have (3.56).

The triangular alpha-admissible notion, defined in [187], is weakened as "triangular α-orbital admissible" by Popescu [230], as follows.

Definition 3.13 [230] For a function $\alpha : X \times X \to [0, \infty)$, a selfmapping $T : X \to X$ is called an α-orbital admissible if

$$\alpha(x, Tx) \geq 1 \Rightarrow \alpha(Tx, T^2 x) \geq 1.$$

In addition, T is said to be a triangular α-orbital admissible if it is both α-orbital admissible and

$$\alpha(x, y) \geq 1 \text{ and } \alpha(y, Ty) \geq 1 \Rightarrow \alpha(x, Ty) \geq 1.$$

Note that each (triangular) α-admissible forms (triangular) α-orbital admissible. The converse is not true: We refer to [183, 230] for distinctive examples.

Theorem 3.49 ([169]) *A continuous selfmapping T, on a complete metric space (X, d) forms both triangular α-orbital admissible and an α-admissible $\mathcal{Z}$-contraction with respect to ζ. If there is $x_0 \in X$ so that $\alpha(x_0, Tx_0) \geq 1$, then, T possesses a fixed point.*

Proof By assumption, we have $x_0 \in X$ so that $\alpha(x_0, Tx_0) \geq 1$. Starting from this point $x_0 \in X$ we construct a recursive sequence $\{x_n\}$ in X by setting $x_{n+1} = Tx_n$ for each $n \geq 0$. Note that in case of the existing an n_0 so that $x_{n_0} = x_{n_0+1}$, we find that $u = x_{n_0}$ is a fixed point of T. Accordingly, we presume that $x_n \neq x_{n+1}$ for each n. Consequently, we deduce

$$d(x_n, x_{n+1}) > 0, \text{ for each } n = 0, 1, \ldots. \tag{3.57}$$

On account of the assumption that T is $\alpha-$admissible, we get

$$\alpha(x_0, x_1) = \alpha(x_0, Tx_0) \geq 1 \text{ implies } \alpha(x_1, x_2) = \alpha(Tx_0, Tx_1) \geq 1.$$

Inductively, we find

$$\alpha(x_n, x_{n+1}) \geq 1, \text{ for each } n = 0, 1, \ldots. \tag{3.58}$$

Taking (3.77) and (3.58) into account, we find

$$\begin{aligned}
0 &\leq \zeta(\alpha(x_n, x_{n-1})d(Tx_n, Tx_{n-1}), d(x_n, x_{n-1})) \\
&= \zeta(\alpha(x_n, x_{n-1})d(x_{n+1}, x_n), d(x_n, x_{n-1})) \\
&< d(x_n, x_{n-1}) - \alpha(x_n, x_{n-1})d(x_{n+1}, x_n),
\end{aligned} \tag{3.59}$$

for each $n \geq 1$. Ergo, we deduce that

$$d(x_n, x_{n+1}) \leq \alpha(x_n, x_{n-1})d(x_n, x_{n+1}) < d(x_n, x_{n-1}) \text{ for each } n = 1, 2, \ldots. \tag{3.60}$$

As a result, $\{d(x_n, x_{n-1})\}$ is non-decreasing sequence that is bounded from below by zero. Accordingly, there is $L \geq 0$ so that $\lim_{n\to\infty} d(x_n, x_{n-1}) = L \geq 0$. We indicate that

$$\lim_{n\to\infty} d(x_n, x_{n-1}) = 0. \tag{3.61}$$

Since the case $L = 0$ terminates the proof, we skip this case and investigate the contrary case, that is, $L > 0$. It is evident that the inequality (3.60) yields

$$\lim_{n\to\infty} \alpha(x_n, x_{n-1})d(x_n, x_{n+1}) = L. \tag{3.62}$$

Setting $s_n = \alpha(x_n, x_{n-1})d(x_n, x_{n+1})$ and $t_n = d(x_n, x_{n-1})$ and together with (ζ_3), we obtain that

$$0 \leq \limsup_{n\to\infty} \zeta(\alpha(x_n, x_{n-1})d(x_{n+1}, x_n), d(x_n, x_{n-1})) < 0 \tag{3.63}$$

a contradiction. Hence, $L = 0$.

Next we indicate that the constructed sequence $\{x_n\}$ is Cauchy via the method of *Reductio ad absurdum*. We presume that the constructed sequence $\{x_n\}$ is not Cauchy. Consequently, there is $\varepsilon > 0$, for each $N \in \mathbb{N}$, there are $n, m \in \mathbb{N}$ in a way that $n > m > N$ and $d(x_m, x_n) > \varepsilon$.

Besides, due to (3.61), there is $n_0 \in \mathbb{N}$ so that

$$d(x_n, x_{n+1}) < \varepsilon \text{ for each } n > n_0. \tag{3.64}$$

Let x_{n_k} and x_{m_k} be two partial subsequences of the constructive sequence x_n in a way that

$$n_0 \leq n_k < m_k < m_{k+1} \text{ with } d(x_{m_k}, x_{n_k}) > \varepsilon \text{ for each } k. \tag{3.65}$$

On the other hand,
$$d(x_{m_k-1}, x_{n_k}) \leq \varepsilon \text{ for each } k, \tag{3.66}$$

where m_k is selected as a least number $m \in \{n_k, n_{k+1}, n_{k+2}, \ldots\}$ so that (3.65) is fulfilled and $n_{k+1} \leq m_k$ for each k.. Notice that the case $n_{k+1} \leq m_k$ is not possible owing to (3.64), (3.65). Accordingly, $n_k + 2 \leq m_k$ for each k. It implies

$$n_{k+1} < m_k < m_{k+1} \text{ for each } k.$$

Combining (3.65), (3.66) together with the triangle inequality, we find

$$\begin{aligned} \varepsilon < d(x_{m_k}, x_{n_k}) &\leq d(x_{m_k}, x_{m_k-1}) + d(x_{m_k-1}, x_{n_k}) \\ &\leq d(x_{m_k}, x_{m_k-1}) + \varepsilon \text{ for each } k. \end{aligned} \tag{3.67}$$

Owing to (3.61), we get
$$\lim_{k\to\infty} d(x_{m_k}, x_{n_k}) = \varepsilon. \tag{3.68}$$

Due to the triangle inequality, together with (3.67) we also find

$$d(x_{m_k}, x_{n_k}) \leq d(x_{m_k}, x_{m_k+1}) + d(x_{m_k+1}, x_{n_k+1}) + d(x_{n_k+1}, x_{n_k}) \text{ for each } k.$$

Analogously, we have

$$d(x_{m_k+1}, x_{n_k+1}) \leq d(x_{m_k+1}, x_{m_k}) + d(x_{m_k}, x_{n_k}) + d(x_{n_k}, x_{n_k+1}) \text{ for each } k.$$

Combining two inequalities above together with (3.61), we find that

$$\lim_{k \to \infty} d(x_{m_k+1}, x_{n_k+1}) = \varepsilon. \tag{3.69}$$

In particular, there is $n_1 \in \mathbb{N}$ so that for each $k \geq n_1$ we have

$$d(x_{m_k}, x_{n_k}) > \frac{\varepsilon}{2} > 0 \text{ and } d(x_{m_k+1}, x_{n_k+1}) > \frac{\varepsilon}{2} > 0. \tag{3.70}$$

Recalling that T is triangular α-orbital admissible, we get

$$\alpha(x_{m_k}, x_{n_k}) \geq 1. \tag{3.71}$$

On account of (3.70) and (3.71) together with the assumption that T is an α-admissible, we find

$$\begin{aligned}
0 &\leq \zeta(\alpha(x_{m_k}, x_{n_k})d(Tx_{m_k}, Tx_{n_k}), d(x_{m_k}, x_{n_k})) \\
&= \zeta(\alpha(x_{m_k}, x_{n_k})d(x_{m_k+1}, x_{n_k+1}), d(x_{m_k}, x_{n_k})) \\
&< d(x_{m_k}, x_{n_k}) - \alpha(x_{m_k}, x_{n_k})d(x_{m_k+1}, x_{n_k+1}),
\end{aligned} \tag{3.72}$$

for each $k \geq n_1$. Consequently, we have

$$0 < d(x_{m_k+1}, x_{n_k+1}) < \alpha(x_{m_k}, x_{n_k})d(x_{m_k+1}, x_{n_k+1}) < d(x_{m_k}, x_{n_k}),$$

for each $k \geq n_1$. From above inequality, together with (3.68) and (3.69), we conclude that $s_n = \alpha(x_{m_k}, x_{n_k})d(x_{m_k+1}, x_{n_k+1}) \to \varepsilon$ as $t_n = d(x_{m_k}, x_{n_k}) \to \varepsilon$. On account of the above observations and regarding the condition (ζ_3), we deduce that

$$0 \leq \limsup_{k \to \infty} \zeta(\alpha(x_{m_k}, x_{n_k})d(x_{m_k+1}, x_{n_k+1}), d(x_{m_k}, x_{n_k})) < 0,$$

a contradiction. Accordingly, the sequence $\{x_n\}$ is Cauchy. Since (X, d) is a complete metric space, there is $x^* \in X$ so that

$$\lim_{n \to \infty} d(x_n, x^*) = 0. \tag{3.73}$$

Recalling that T is continuous, the expression (3.73) yields

$$\lim_{n \to \infty} d(x_{n+1}, Tx^*) = \lim_{n \to \infty} d(Tx_n, Tx^*) = 0. \tag{3.74}$$

From (3.73) to (3.74), we deduce, from the uniqueness of the limit, that x^* is a fixed point of T.   □

Theorem 3.50 *A continuous selfmapping* T, *on a complete metric space* (X, d) *is regular, triangular* α-*orbital admissible and an* α-*admissible* $\mathcal{Z}$-*contraction with respect to* ζ. *If there is* $x_0 \in X$ *so that* $\alpha(x_0, Tx_0) \geq 1$, *then,* T *possesses a fixed point.*

Proof Verbatim the initial lines of the proof of Theorem 3.49, we derive a converges sequence $\{x_n\}$ defined by $x_{n+1} = Tx_n$ for each $n \geq 0$ with a limit x^*. Since it is regular, the expression (3.58) yields that there is a subsequence $\{x_{n(k)}\}$ of $\{x_n\}$ so that $\alpha(x_{n(k)}, x^*) \geq 1$ for each k. Employing (3.77), for each k, we obtain

$$
\begin{aligned}
0 &\leq \zeta(\alpha(x_{n(k)}, x^*)d(Tx_{n(k)}, Tx^*), d(x_{n(k)}, x^*)) \\
&= \zeta(\alpha(x_{n(k)}, x^*)d(x_{n(k)+1}, Tx^*), d(x_{n(k)}, x^*)) \\
&< d(x_{n(k)}, x^*) - \alpha(x_{n(k)}, x^*)d(x_{n(k)+1}, Tx^*),
\end{aligned}
\tag{3.75}
$$

that is equivalent to

$$
d(x_{n(k)+1}, Tx^*) = d(Tx_{n(k)}, Tx^*) \leq \alpha(x_{n(k)}, x^*)d(Tx_{n(k)}, Tx^*) \leq d(x_{n(k)}, x^*).
\tag{3.76}
$$

Setting $k \to \infty$ in the equality above , we find $d(x^*, Tx^*) = 0$, that is, $x^* = Tx^*$. □

Notice that in Theorems 3.49 and 3.50, founded fixed point is not necessarily unique. For its uniqueness, we need an additional condition:

(U) For each $x, y \in \text{Fix}(T)$, we have $\alpha(x, y) \geq 1$.

The uniqueness of the fixed point in Theorems 3.49 and 3.50 is settled as follows:

Theorem 3.51 *If we presume the hypotheses* (U) *in addition to the statement of Theorem 3.49 (resp. Theorem 3.50), then we conclude that fixed point,* x^*, *of* T *is the unique.*

3.8.1 Immediate Consequences

On account of Example 3.7.3, we can easily list immediate consequence Theorem 3.51.
  In particular, for a (c)-comparison function ψ, we construct a new function

$$
\zeta_E(t, s) = \psi(s) - t \qquad \text{for each } s, t \in [0, \infty),
$$

which forms a simulation function (cf. Example 3.7.3 (v)).
  It is an evident to deduced that the main result of Samet et al. [254] can be stated as a consequence of Theorem 3.51:

Theorem 3.52 *Theorem 3.51 implies Theorem 3.43.*

Proof If we set $\zeta_E(t, s) = \psi(s) - t$ for each $s, t \in [0, \infty)$ in Theorem 3.51, then

$$\alpha(x, y)d(Tx, Ty) \leq \psi(d(x, y)), \text{ for each } x, y \in X.$$

The rest is evident, so we skip. ☐

In other words, Theorem 3.52 indicates that all obtained corollaries of Theorem 3.43 are natural and immediate consequences of Theorem 3.51.

In a similar way, we note that the main result of Khojasteh et al. [194] is also consequences of Theorem 3.51.

Theorem 3.53 *Theorem 3.51 yields Theorem 3.47.*

Proof It is sufficient to set $\alpha(x, y) = 1$ for each $x, y \in X$. ☐

In other words, Theorem 3.53 indicates that all obtained corollaries of Theorem 3.47 are natural and immediate consequences of Theorem 3.51.

After the discussion on the fundamental and initial results in this direction, we consider a more general case by the following definition:

Definition 3.14 For a self-mapping T, defined on a metric space (X, d), suppose that there are auxiliary functions $\zeta \in \mathcal{Z}$ and $\alpha : X \times X \to [0, \infty)$ so that

$$\zeta(\alpha(x, y)d(Tx, Ty), K(x, y)) \geq 0 \qquad \text{for each } x, y \in X. \tag{3.77}$$

Then, T is called a generalized K-type α-admissible $\mathcal{Z}$-*contraction* with respect to ζ, where $K(x, y)$ is considered as one of the expressions in the following list ($\forall x, y \in X$):

$$= \max \left\{ d(x, y), d(x, Tx), d(y, Ty), \frac{d(x, Ty) + d(y, Tx)}{2} \right\};$$
$$= \max \{ d(x, Tx), d(y, Ty) \};$$
$$= \max \{ d(x, y), d(x, Tx), d(y, Ty) \};$$
$$= \max \left\{ d(x, y), \frac{d(x, Ty) + d(y, Tx)}{2} \right\};$$
$$= \max \left\{ d(x, y), \frac{[1 + d(x, Tx)]d(y, Ty)}{1 + d(x, y)} \right\};$$
there are $\alpha, \beta, \gamma \in [0, 1)$ with $\alpha + \beta + \gamma < 1$ so that
$$= \alpha d(x, y) + \beta d(x, Ty) + \gamma d(y, Tx), \ .$$

Remark The function $K(x, y)$ can be extended in further distinct ways more. In Definition 3.14, we restrict ourselves to the fundamental contraction forms. The following theorem can be considered as an example of this thought.

Theorem 3.54 *A selfmapping T on a complete metric space* (X, d) *is a triangular α-orbital admissible and forms a generalized K-type α-admissible $\mathcal{Z}$-contraction with respect to ζ. We presume that there is $x_0 \in X$ so that $\alpha(x_0, Tx_0) \geq 1$. If a mapping T is continuous, or (X, d) is regular, then T possesses a unique fixed point.*

The proof is slight modification of the proofs of Theorem 3.49 and Theorem 3.50.

Theorem 3.55 *If we presume the hypotheses (U) in addition to the statement of Theorem 3.54, then we conclude that fixed point, x^*, of T is the unique.*

The proof is slight modification of the proofs of Theorem 3.51.

3.9   Fixed Points via Two Metrics

Theorem 3.56 (Maia [208]) *Let T be a self-mapping on X. Let (X, d) and (X, δ) be two metric spaces and T be a self-mapping on X. Assume that the following conditions are satisfied:*

(M1) $d(x, y) \leq \delta(x, y)$ *for each* $x, y \in X$,
(M2) X *is complete with respect to* d,
(M3) T *is continuous with respect to* d,
(M4) *there is a constant* $k \in [0, 1)$ *so that*

$$\delta(Tx, Ty) \leq k\delta(x, y),$$

for each $x, y \in X$.

Then T has a unique fixed point in X.

Part II
Fixed Point Theorems on Various Metric Spaces

In this part, the aim is to investigate the existence and uniqueness of the fixed points in the setting of new abstract spaces.

Generalization of Metric Spaces   4

In this chapter, we discuss some of the interesting generalizations and extensions of the notion of the metric. Roughly speaking, the notion of metric can be considered as an axiomatic form of the "distance". For this reason, the metric and all generalizations and extensions of the metric can be called "distance function". One of the well-known examples of the metric is due to Euclid which is known as Euclidean metric.

4.1   *b*-Metric Space

In this section, we consider *b*-metric which has attracted the attention of researchers in the last few decades. Indeed, the notion of *b*-metric has been considered, independently and separately, by many authors, see e.g. [57, 73, 103]. In some sources, *b*-metric appeared in the name of "quasi-metric" see e.g. [58–60]. In this manuscript, we shall use *b*-metric since "quasi-metric" has been used for different generalization of the metric. The concept of *b*-metric has been known and popular after the paper Czerwik [103] who introduce the notion of *b*-metric by relaxing the axiom of triangle inequality in the assumption of the concept of the metric. and introduce Indeed, this structure was already proposed early by some other authors, like Bakhtin [57], Bourbaki [73] and so on. Czerwik [103] characterized the analog of Banach contraction mapping principle in the context of *b*-metric spaces. After this pioneer result of Czerwik [103] on *b*-metric, a vast number of authors have published a huge number of papers in this direction (see e.g. [5, 9, 36, 37, 70–78] and related references therein).

For the sake of completeness, we recollect the notion as follows:

Definition 4.1 (*Czerwik* [103]) Let X be a nonempty set and $d : X \times X \to [0, \infty)$ be a function satisfying the following conditions:

© The Author(s), under exclusive license to Springer Nature Switzerland AG 2022   73
E. Karapınar and R. P. Agarwal, *Fixed Point Theory in Generalized Metric Spaces*,
Synthesis Lectures on Mathematics & Statistics,
https://doi.org/10.1007/978-3-031-14969-6_4

(b1) $d(x, y) = 0$ if and only if $x = y$; (self-distance).
(b2) $d(x, y) = d(y, x)$; (symmetry).
(b3) $d(x, y) \leq s[d(x, z) + d(z, y)]$, (extended-triangle inequality).

for all $x, y, z \in X$, where $s \geq 1$. A function d is called b-metric whenever it satisfies all the axioms above. In addition, the triple (X, d, s) is called b-metric space.

A b-metric coincides with the corresponding standard metric if $s = 1$.
   The following is the fundamental and interesting example of b-metric:

Example see e.g. [70–75]

Set $Y = \{x, y, z\}$ and $X = Y \cup \mathbb{N}$. Define a function $d : X \times X \to [0, \infty)$ so that

$$d(y, z) = d(z, y) = \Delta,$$
$$d(x, y) = d(y, x) = d(x, z) = d(z, x) = 1,$$
$$d(p, q) = \left| \frac{1}{p} - \frac{1}{q} \right| \text{ for all } p, q \in \mathbb{N}, d(x, x) = d(y, y) = d(z, z) = 0,$$

with $\Delta \in [2, \infty)$. Then, we observe

$$d(x, y) \leq \frac{\Delta}{2} [d(x, z) + d(z, y)], \text{ for } x, y, z \in Y.$$

Accordingly, (X, d) forms a b-metric space. It should be underlined that incase $\Delta > 2$ the triangle inequality does not hold. Accordingly, (X, d) does not form a metric.

For the immediate examples on b-metrics we state the followings.

Example see e.g. [70–75]

Set $d : \mathbb{R} \times \mathbb{R} \to [0, \infty)$ with the rule

$$d(x, y) = |x - y|^p \tag{4.1}$$

for $p > 1$. Ergo, d forms a b-metric on $\mathbb{R}$. Indeed, the first two conditions are trivial. For the next one, since

$$|x - y|^p \leq 2^{p-1}[|x - z|^p + |z - y|^p],$$

the third condition holds with $s = 2^{p-1}$. Consequently, $(\mathbb{R}, d, s)$ forms a b-metric space with $s = 2^{p-1}$.

Example see e.g. [70–75]

For $p \in (0, 1)$, let

$$X = l_p(\mathbb{R}) = \left\{ x = \{x_n\} \subset \mathbb{R} : \sum_{n=1}^{\infty} |x_n|^p < \infty \right\}.$$

Set $d : X \times X \to [0, \infty)$ in a way that

$$d(x, y) = \left(\sum_{n=1}^{\infty} |x_n - y_n|^p \right)^{1/p}.$$

Accordingly, (X, d, s) forms a *b*-metric space with $s = 2^{1/p}$.

Example [70–75]

Let P be a cone in Banach space E with $int(P) \neq \emptyset$ and 0_E be the zero vector of E. Suppose that the relation $\preceq$ forms a partial ordering with respect to P. For a nonempty set, the mapping $d : X \times X \to E$ is called cone metric on X if it fulfills:

(M1) $0 \preceq d(x, y)$ for all $x, y \in X$,
(M2) $d(x, y) = 0$ if and only if $x = y$,
(M3) $d(x, y) = d(y, x)$ for all $x, y \in X$,
(M4) $d(x, y) \preceq d(x, z) + d(z, y)$, for all $x, y \in X$.

On the other hand, (X, d) forms a cone metric space.

Let K be the normality constant of the normal cone P in a Banach space E. Set $D : X \times X \to [0, \infty)$ with $D(x, y) = \|d(x, y)\|$, where $d : X \times X \to E$ is a cone metric space. It is straightforward to see that (X, D, s) forms a *b*-metric space with $s := K \geq 1$.

Remark 4.1 It is worth mentioning that, as a function, (*b*)-metric is not necessarily continuous, see e.g. [?].

4.2   Partial Metric Spaces

In mathematics, we always aim to get a more general case. The researchers' goal is to get a well-known result in the setting of the most general case/structure. From the point of view of a mathematician, dealing with infinite sets and sequences is more interesting than the corresponding finite cases. On the other hand, from a computer scientist's point of view, it is just the converse. Computer scientists aim to reduce the infinite possibility into the finite case. More precisely, in the domain theory of computer science, the most crucial question is how to terminate the software program when dealing with infinite sequences. Based on the motivation above, Matthews [210] proposed a new metric structure, partial metric space. To make the idea of the partial metric, the well-known Baire metric can be helpful. By involving the finite sequences in the setting of the standard Baire metric, we obtain one of the most exciting examples of partial metric [210], as follows:

A mapping $d : \omega_i \times \omega_i : [0, \infty)$, defined by

$$d(x, y) = 2^{-\sup\{n|\forall i < n \text{ such that } x_i = y_i\}}, \tag{4.2}$$

for all $x, y \in \omega_i$, satisfies all hypotheses of standard metric, where ω_i denotes the set of all infinite sequences. Here, the pair (ω_i, d) is called as Baire metric space. On the other hand, we change the domain of the function ω_i with the set of all both finite ω_f and infinite sequences ω_i, we observe that the self-distance is not necessarily zero. Notice that the extended function $p : (\omega_i \cup \omega_f) \times (\omega \cup \omega_f) \to [0, \infty)$, defined by

$$p(x, y) = 2^{-\sup\{n|\forall i < n \text{ such that } x_i = y_i\}}, \tag{4.3}$$

for all $x, y \in \omega_i \cup \omega_f$, does not form a metric, since $p(x, x) \neq 0$ for all $x \in \omega_f$. In particular, for $x \in \omega_f$ with $x = (x_1, x_2, \cdots, x_{19})$. thus, $p(x, x) = \frac{1}{2^{19}} \neq 0$.

Motivated the discussion above, Matthews [210], proposed a new notion, partial metric, as follows.

Definition 4.2 Let X be a nonempty set. A function $p : X \times X \to [0, \infty)$ is called a partial metric on X if it satisfies the properties

(p_1) $p(x, x) = p(x, y)$ and $p(x, x) = p(y, y)$ implies $x = y$,
(p_2) $p(x, x) \leq p(x, y)$,
(p_3) $p(x, y) = p(y, x)$,
(p_4) $p(x, z) + p(y, y) \leq p(x, y) + p(y, z)$, (weakened-triangle inequality)

for all $x, y, z \in X$. The pair (X, p) is referred to as a partial metric space (see [210, 216] for more details on the subject).

It is clear that the pair $(\omega \cup \omega_f, p)$ provides all hypotheses of a partial metric space. So, modified Baire metric is the first example of the partial metric. As second example, the operator of maximum on the nonnegative real numbers, denoted by $\mathbb{R}_0^+$, fulfills the hypotheses of partial metric space.

Example [See e.g. [210, 211]]

Let $p : \mathbb{R}_0^+ \times \mathbb{R}_0^+ \longrightarrow \mathbb{R}_0^+$ is defined by $p(x, y) = \max\{x, y\}$ for each $x, y \in X$. Consequently, we have $\max\{x, x\} = x \geq 0$ and $\max\{x, x\} = 0$ if and only if $x = 0$. Thus, (X, p) forms a partial metric space.

Example [See e.g [154, 266]]

Let (X, d) be standard metric space and (X, p) be a partial metric space. We presume that $a \geq 0$ and $\omega : X \to \mathbb{R}_0^+$ is an arbitrary function. Define the functions $\rho_i : X \times X \longrightarrow \mathbb{R}_0^+$ $(i \in \{1, 2, 3\})$ as follow by

$$\rho_1(x, y) = p(x, y) + a;$$
$$\rho_2(x, y) = p(x, y) + \max\{\omega(x), \omega(y)\};$$
$$\rho_3(x, y) = p(x, y) + p(x, y).$$

These functions, ρ_1, ρ_2, ρ_3, are partial metrics on X.

Example [See [216]]

Let $X = \{[a, b] : a, b, \in \mathbb{R}, \ a \leq b\}$ and define $p([a, b], [c, d]) = \max\{b, d\} - \min\{a, c\}$. Clearly, (X, p) forms a partial metric space.

Example

For a set $X := [0, \frac{1}{2}] \cup [4, 6]$, we consider $p : X \times X \to [0, \infty)$ which is defined by

$$p(x, y) = \begin{cases} |x - y| \text{ if } \{x, y\} \subset [0, \frac{1}{2}], \\ \max\{x, y\} \text{ if } \{x, y\} \cap [4, 6] \neq \emptyset. \end{cases}$$

Clearly, p forms a partial metric and hence (X, p) forms a partial metric space.

For the fundamental topological properties of a partial metric space (X, p), we define open ball as follows:

$$B_p(x, \gamma) = \{y \in X : p(x, y) < p(x, x) + \gamma\} \text{ for all } x \in X \text{ and } \gamma > 0.$$

Further, $\{B_p(x, \gamma) : x \in X, \gamma > 0\}$, the family of open p-balls, forms a base for the topology of (X, p). In addition, the family of open p-balls generates a T_0 topology τ_p on X.

For a partial metric space (X, p), set $d_p : X \times X \to \mathbb{R}_0^+$ and $d_m^p : X \times X \to \mathbb{R}_0^+$ as follows

$$d_p(x, y) = 2p(x, y) - p(x, x) - p(y, y), \tag{4.4}$$

and

$$\begin{aligned} d_m^p(x, y) &= \max\{p(x, y) - p(x, x), p(x, y) - p(y, y)\} \\ &= p(x, y) - \min\{p(x, x), p(y, y)\}, \end{aligned} \tag{4.5}$$

Clearly, it is a (standard) metric on X. In addition, $p_0 : X \times X \to \mathbb{R}_0^+$ forms a metric with the following definition (see e.g. [29, 133])

$$\begin{cases} p_0(x, x) = 0, & \text{for all } x \in X, \\ p_0(x, y) = p(x, y), & \text{for all } x \neq y. \end{cases} \tag{4.6}$$

Note that any partial metric p, we have $\tau_p \subseteq \tau_{d_p} = \tau_{d_p^m} \subseteq \tau_{p_0}$.

Definition 4.3 ([34, 216]) For a partial metric space (X, p), we have

(1) a sequence $\{x_n\}$ in (X, p) converges to $x \in X$ if $p(x, x) = \lim_{n \to \infty} p(x, x_n)$;
(2) a sequence $\{x_n\}$ in (X, p) is called a Cauchy sequence if $\lim_{m,n \to \infty} p(x_m, x_n)$ exists (and is finite);
(3) (X, p) is said to be complete if every Cauchy sequence $\{x_n\}$ in X converges, with respect to τ_p, to a point $x \in X$ such that $p(x, x) = \lim_{m,n \to \infty} p(x_m, x_n)$;
(4) a subset A of a partial metric space (X, p) is closed in (X, p) if it contains its limit points, that is, if a sequence $\{x_n\}$ in A converges to some $x \in X$, then $x \in A$.
(5) a subset A of a partial metric space (X, p) is bounded in (X, p) if there exist $x_0 \in X$ and $M \in \mathbb{R}$ such that for all $a \in A$, we have $a \in B_p(x_0, M)$, that is, $p(x_0, a) < p(a, a) + M$.

Definition 4.4 For a partial metric space (X, p), a self-mapping T on X is called continuous, if for each sequence $\{x_n\}$ in X converges to $u \in X$, that is,

$$\lim_{n \to \infty} p(x_n, u) = \lim_{n \to \infty} p(x_n, x_{n+k}) = p(u, u) \tag{4.7}$$

provides

$$\lim_{n \to \infty} p(Tx_n, Tu) = \lim_{n \to \infty} p(Tx_n, Tx_{n+k}) = P(Tu, Tu). \tag{4.8}$$

It is worth mentioning that the equality (4.8) can be stated as

$$\lim_{n\to\infty} p(Tx_n, Tu) = \lim_{n\to\infty} p(x_{n+1}, Tu) = \lim_{n\to\infty} p(x_{n+1}, x_{n+k+1}) = p(u, u)$$
$$= \lim_{n\to\infty} p(Tx_n, Tx_{n+k}) = P(Tu, Tu). \tag{4.9}$$

Remark 4.2 The limit in a partial metric space may not be unique. For example, consider the sequence $\{\frac{1}{n^2+n}\}_{n\in\mathbb{N}}$ in the partial metric space (X, p) where $p(x, y) = \max\{x, y\}$. Note

$$p(1, 1) = \lim_{n\to\infty} p\left(1, \frac{1}{n^2+n}\right) \quad \text{and} \quad p(2, 2) = \lim_{n\to\infty} p\left(2, \frac{1}{n^2+n}\right).$$

The following lemma is crucial to see the connection between the metric space and partial metric space.

Lemma 4.1 ([216, 222])

(i) $\{x_n\}$ is a Cauchy sequence in a partial metric space (X, p) if and only if it is a Cauchy sequence in the metric space (X, d_p);

(ii) A partial metric space (X, p) is complete if and only if the metric space (X, d_p) is complete. Furthermore, $\lim_{n\to\infty} d_p(x_n, x) = 0$ if and only if $p(x, x) = \lim_{n\to\infty} p(x_n, x) = \lim_{n\to\infty} p(x_n, x_m)$.

(iii) If $\{x_n\}$ is a convergent sequence in (X, d_p), then it is a convergent sequence in the partial metric space (X, p).

Lemma 4.2 (See e.g. [2]) Let $\{x_n\}$ and $\{y_n\}$ be two sequences in a partial metric space X such that

$$\lim_{n\to\infty} p(x_n, x) = \lim_{n\to\infty} p(x_n, x_n) = p(x, x), \tag{4.10}$$

and

$$\lim_{n\to\infty} p(y_n, y) = \lim_{n\to\infty} p(y_n, y_n) = p(y, y), \tag{4.11}$$

then $\lim_{n\to\infty} p(x_n, y_n) = p(x, y)$. In particular, $\lim_{n\to\infty} p(x_n, z) = p(x, z)$ for every $z \in X$.

Lemma 4.3 (See e.g. [157]) Let (X, p) be a partial metric space. Then

(A) If $p(x, y) = 0$ then $x = y$,
(B) If $x \neq y$, then $p(x, y) > 0$.

Lemma 4.4 ([176]) Let (X, p) be a partial metric space. A sequence $\{x_n\}_{n\in\mathbb{N}}$ in X is a Cauchy sequence in (X, p) if and only if it satisfies the following condition:

(*) for each $\varepsilon > 0$ there is $n_0 \in \mathbb{N}$ such that $p(x_n, x_m) - p(x_n, x_n) < \varepsilon$ whenever $n_0 \leq n \leq m$.

Lemma 4.5 (See e.g. [157]) *Assume $x_n \to z$ as $n \to \infty$ in a partial metric space(X, p) such that $p(z, z) = 0$. Then $\lim_{n \to \infty} p(x_n, y) = p(z, y)$ for every $y \in X$.*

4.3   On Further Extension of the Metric Notion: Overview

In this short section, we shall consider all interesting conditions for a distance function $d :$ $X \times X \to [0, \infty)$ where the set X presumed nonempty. Some of them are already discussed in the metric section. For the sake of completeness, we repeat here. Consider the following assumptions for every $x, y, z \in X$:

(d_1) $d(x, x) = 0$;
(d_2) $d(x, y) = d(y, x) = 0 \implies x = y$;
(d_3) $d(x, y) = d(y, x)$;
(d_4) $d(x, z) \le d(x, y) + d(y, z)$;
(d_{4a}) $d(x, z) \le d(x, y) + d(y, z) - d(z, z)$;
(d_{4b}) $d(x, z) \le d(x, y) + d(y, w) + d(w, z)$ distinct y, w;
(d_{4c}) $d(x, z) \le s[d(x, y) + d(y, z)]$ for some $s \ge 1$;
(d_{4d}) $d(x, z) \le \max\{d(x, y), d(y, z)\}$;
(d_5) $d(x, x) \le d(x, y)$.

Regarding the listed conditions above, we can give the following chart to indicate the well-known generalizations and extensions of the notion of the metric.

Metric	$(d_1) - (d_4)$
Bianciari-distance [47, 48, 54, 64, 68, 81, 82, 105, 113, 115, 147, 151, 160, 162, 168, 184–186, 195, 257]	$(d_1) - (d_3), (d_{4b})$
Dislocated-metric [133]	$(d_2) - (d_4)$
Quasi-metric [24, 35, 35, 67, 91, 132]	$(d_1), (d_2), (d_4)$
Semi-metric [279]	$(d_1), (d_2), (d_3)$
partial metric [2, 157, 210, 211]	$(d_2), (d_3), (d_{4a})$ and (d_5)
b-metric [5, 9, 23, 36, 37, 70–72, 74, 75, 158]	$(d_1) - (d_3), (d_{4c})$
ultra metric	$(d_1) - (d_3), (d_{4d})$
metric-like [12, 19, 127]	$(d_2) - (d_4)$

It is worth mentioning that there are several more distance functions that appeared as generalizations and expansions of the metric notion. Besides the listed notions above, we can count, for example, fuzzy metric [198], probabilistic metric [124, 131], Menger Space [214], multiplicative metric, 2-metric, D-metric, G-metric [13, 14, 219], S-metric, cone metric (Banach-valued metric) [62, 63, 106, 134, 143, 149, 172], complex-valued metric, quaternion-valued metric, C*-algebra-valued metric, modular metric and so on. We refer to the following papers [8, 25, 107, 108, 150, 158, 256] for the explicit definitions of some of the mentioned notions with strong remarks.

In this manuscript, we restrict ourselves to only two notions: partial metric and b-metric. These two notions are the real generalization of the metric notion and very interesting in the point view of applications. For these reasons, both of them have been investigated heavily by researchers and a remarkable number of papers have appeared on the topics in the corresponding literature.

Fixed Point Theorems on b-Metric Spaces

In this chapter, we collect some important fixed point theorems. Fixed point theorems in b-metric spaces have been studied by many author, e.g. [11, 12, 20, 23, 36, 37, 45, 57, 59, 71, 72, 74, 76, 77, 79, 117, 119, 161, 182, 217].

5.1 Fixed Point Theorems via Admissible Mappings

We start this section by recalling the crucial concept of the (b)-comparison function which is a natural extension of (c)-comparison function. In the metric fixed point literature, there are very interesting fixed point theorems that are obtained by using the (c)-comparison function. In order to get the characterization of such theorems in the framework of b-metric spaces, Berinde [59] proposed the notion of (b)-comparison function, as follows:

Definition 5.1 (*Berinde* [59]) For a real number $s \geq 1$, a mapping $\varphi : [0, \infty) \rightarrow [0, \infty)$ is called a (b)-comparison function if

(1) it is monotone increasing;

(2) there exist $k_0 \in \mathbb{N}$, $a \in (0, 1)$ and a convergent series of nonnegative terms $\sum_{k=1}^{\infty} v_k$ such that $s^{k+1} \varphi^{k+1}(t) \leq a s^k \varphi^k(t) + v_k$, for $k \geq k_0$ and any $t \in [0, \infty)$.

We reserve the symbol Ψ_b to demote the class of (b)-comparison function $\varphi : [0, \infty) \rightarrow [0, \infty)$. It is trivial to conclude that (b)-comparison function coincides with (c)-comparison function when $s = 1$.

For the (b)-comparison function, we have the following interesting observations.

E. Karapınar and R. P. Agarwal, *Fixed Point Theory in Generalized Metric Spaces*, Synthesis Lectures on Mathematics & Statistics, https://doi.org/10.1007/978-3-031-14969-6_5

Lemma 5.1 (Berinde [58]) *For a (b)-comparison function $\varphi : [0, \infty) \to [0, \infty)$ we have*

(1) the series $\displaystyle\sum_{k=0}^{\infty} s^k \varphi^k(t)$ converges for any $t \in [0, +\infty)$;

(2) $b_s : [0, \infty) \to [0, \infty)$ is expressed as $b_s(t) = \displaystyle\sum_{k=0}^{\infty} s^k \varphi^k(t)$, $t \in [0, \infty)$, is continuous

   at 0 and increasing.

Remark 5.1 (*see e.g.* [135]) Since (b)-comparison function is also comparison function by Lemmas 5.1, and 2.4, we deduce that any (b)-comparison function ϕ satisfies $\phi(t) < t$, for any $t > 0$.

In what follows, we recall the notion of α-ψ-contractive mapping of type-(b).

Definition 5.2 Let T be a selfmapping on a b-metric space (X, d, s) with constant $s > 1$. Then, T is called an α-ψ-contractive mapping of type-(b) if there are two functions $\alpha : X \times X \to [0, \infty)$ and $\psi \in \Psi_b$ in a way that

$$\alpha(x, y)d(T(x), T(y)) \leq \psi(d(x, y)), \text{ for each } x, y \in X. \tag{5.1}$$

The following is an interesting fixed point theorem via the notion of α-ψ-contractive mapping of type-(b).

Theorem 5.1 *Let T be a selfmapping on a b-metric space (X, d, s) with constant $s > 1$. We presume that T is continuous, α-admissible and forms an α-ψ-contractive mapping of type-(b) satisfying the following conditions: If there is $x_0 \in X$ such that $\alpha(x_0, T(x_0)) \geq 1$, then, T possesses a fixed point.*

Proof By assumption, we have an initial point $x_0 \in X$ so that $\alpha(x_0, T(x_0)) \geq 1$. We initiate a sequence $\{x_n\}$ in X by the definition

$$x_{n+1} = T(x_n), \text{ for each } n \in \mathbb{N}_0.$$

Incase of having the equality $x_n = x_{n+1}$ for some $n \in \mathbb{N}_0$, then $x^* = x_n$ forms a fixed point for T that finalize the proof. Accordingly, we assume, on the contrary, that

$$x_n \neq x_{n+1} \text{ for each } n \in \mathbb{N}_0. \tag{5.2}$$

Regarding that T is α-orbital admissible, we find

$$\alpha(x_0, x_1) = \alpha(x_0, T(x_0)) \geq 1 \text{ implies that } \alpha(T(x_0), T(x_1)) = \alpha(x_1, x_2) \geq 1.$$

Recursively, we obtain

$$\alpha(x_n, x_{n+1}) \geq 1, \text{ for each } n \in \mathbb{N}_0. \tag{5.3}$$

Employing the inequality (5.1) with $y = x_n$ and $x = x_{n-1}$, and keeping (5.3) in mind, we find

$$d(x_n, x_{n+1}) = d(T(x_{n-1}), T(x_n)) \leq \alpha(x_{n-1}, x_n)d(T(x_{n-1}), T(x_n))$$
$$\leq \psi(d(x_{n-1}, x_n)).$$

Iteratively, we obtain

$$d(x_n, x_{n+1}) \leq \psi^n(d(x_0, x_1)), \text{ for each } n \in \mathbb{N}_0. \tag{5.4}$$

Taking (5.4) into account, together with the extended-triangle inequality, we derive that

$$d(x_n, x_{n+p}) \leq sd(x_n, x_{n+1}) + s^2 d(x_{n+1}, x_{n+2}) + \cdots + s^{p-2} d(x_{n+p-3}, x_{n+p-2})$$

$$+ s^{p-1} d(x_{n+p-2}, x_{n+p-1}) + s^p d(x_{n+p-1}, x_{n+p})$$

$$\leq s\psi^n(d(x_0, x_1)) + s^2 \psi^{n+1}(d(x_0, x_1)) + \cdots + s^{p-2} \psi^{n+p-3}(d(x_0, x_1))$$

$$+ s^{p-1} \psi^{n+p-2}(d(x_0, x_1)) + s^{p-1} \psi^{n+p-1}(d(x_0, x_1))$$

$$= \frac{1}{s^{n-1}}[s^n \psi^n(d(x_0, x_1)) + s^{n+1} \psi^{n+1}(d(x_0, x_1)) + \cdots + s^{n+p-2} \psi^{n+p-2}(d(x_0, x_1))$$

$$+ s^{n+p-1} \psi^{n+p-1}(d(x_0, x_1))],$$

for each $p \geq 1$.

Now, set $S_n = \sum_{k=0}^{n} s^k \psi^k(d(x_0, x_1))$, $n \geq 1$ we get

$$d(x_n, x_{n+p}) \leq \frac{1}{s^{n-1}}[S_{n+p-1} - S_{n-1}], \ n \geq 1, \ p \geq 1. \tag{5.5}$$

Owing to Lemma 5.1 together with the assumption (5.2), we deduce that the series $\sum_{k=0}^{n} s^k \psi^k(d(x_0, x_1))$ is convergent. Consequently, there is $S = \lim_{n \to \infty} S_n \in [0, \infty)$. Since $s \geq 1$, by (5.5), we derive that $\{x_n\}_{n \geq 0}$ forms a Cauchy sequence in the framework of b-metric space (X, d). Regarding that (X, d, s) is complete, there is $x^* \in X$ so that $x_n \to x^*$ as $n \to \infty$. Due to the continuity of T, we find $x_{n+1} = T(x_n) \to T(x^*)$ as $n \to \infty$. Owing to the uniqueness of the limit, we deduce $x^* = T(x^*)$. □

Example

Let $X = (-\infty, 2]$, the b-metric $d : X \times X \to [0, \infty)$, where $d(x, y) = (x - y)^2$, and let $T : X \to X$, with $Tx = \begin{cases} x \cdot e^x, & \text{if } x < 0 \\ \frac{x}{2}, & \text{if } x \in [0, 2] \end{cases}$, be a continuous mapping. Choosing $\alpha :$ $X \times X \to [0, \infty)$ defined as $\alpha(x, y) = \begin{cases} 1, & \text{if } x, y \in [0, 2] \\ 0, & \text{otherwise} \end{cases}$ and $\psi \in \Psi_b$, $\psi(t) = \frac{t}{3}$, we have for any $x, y \in [0, 2]$

$$\alpha(x, y)d(Tx, Ty) = \frac{(x - y)^2}{4} \leq \frac{(x - y)^2}{3} = \psi(d(x, y)).$$

Therefore, (since it is simple to note that due to the way the function α was chosen, the other cases are not interesting) all the assumption of Theorem 5.1. are satisfied, and the mapping T has a fixed point, that is $x = 0$.

The aim of the next theorem is to avoid the necessity of continuity of T. We indicate that it is possible by replacing the continuity of T with a regularity of the space.

Theorem 5.2 *Let T be a selfmapping on a b-metric space (X, d, s) with constant $s > 1$. We presume that T is α-admissible and forms an α-ψ-contractive mapping of type-(b) satisfying the statement: If (X, d, s) is regular and if there is $x_0 \in X$ such that $\alpha(x_0, T(x_0)) \geq 1$, then, T possesses a fixed point.*

Proof By verbatim of the initial lines of the proof of Theorem 5.1, we derive that the constructed sequence $\{x_n\}$ is convergent to a point $x^* \in X$. On account of the regularity of the space and (5.3), we find

$$\alpha(x_n, x^*) \geq 1, \text{ for each } n \in \mathbb{N}. \tag{5.6}$$

Due to the extended-triangle inequality, (5.1) and (5.6), we obtain

$$\begin{aligned} d(T(x^*), x^*) &\leq s[d(T(x^*), T(x_n)) + d(x_{n+1}, x^*)] \\ &\leq s[d(T(x^*), x_{n+1}) + d(x_{n+1}, x^*)] \\ &\leq s[\alpha(x_n, x^*)d(T(x^*), T(x_n)) + d(x_{n+1}, x^*)] \\ &\leq s[\psi(d(x_n, x^*)) + d(x_{n+1}, x^*)]. \end{aligned}$$

Setting $n \to \infty$ and keeping in mind that ψ is continuous at the point $t = 0$, we conclude $d(T(x^*), x^*) = 0$ and hence $x^* = T(x^*)$. □

To assure the uniqueness of the fixed point, we need an additional hypothesis, as follows:

$(H):$ for each $x, y \in X$, there is $z \in X$ such that $\alpha(x, z) \geq 1$ and $\alpha(y, z) \geq 1$.

Theorem 5.3 *Adding condition (H) to the hypotheses of Theorem 5.1 (resp. Theorem 5.2) we deduce uniqueness of the fixed point of T.*

Proof We presume that x^* and y^* are two distinct fixed point of T. Taking (H) into account, there is $z \in X$ so that

$$\alpha(x^*, z) \geq 1 \quad \text{and} \quad \alpha(y^*, z) \geq 1. \tag{5.7}$$

Recalling that T is α-orbital admissible, the expression (5.7) yields

$$\alpha(x^*, T^n(z)) \geq 1 \quad \text{and} \quad \alpha(y^*, T^n(z)) \geq 1. \tag{5.8}$$

Using (5.8) and (5.1), we have

$$d(x^*, T^n(z)) = d(T(x^*), T(T^{n-1}(z))) \leq \alpha(x^*, T^{n-1}(z))d(T(x^*), T(T^{n-1}(z)))$$

$$\leq \psi(d(x^*, T^{n-1}(z))).$$

Accordingly, we find

$$d(x^*, T^n(z)) \leq \psi^n(d(x^*, z)), \text{ for each } n \in \mathbb{N}.$$

Then, letting $n \to \infty$, we have

$$T^n(z) \to x^*. \tag{5.9}$$

Similarly, using (5.8) and (5.1), we get

$$T^n(z) \to y^* \text{ as } n \to \infty. \tag{5.10}$$

Using (5.9) and (5.10), the uniqueness of the limit yields that $x^* = y^*$. This completes the proof. □

5.2 Nonunique Fixed Points Theorems

In this section we indicate how nonunique fixed point theorems are extended in the framework of b-metric spaces. First we underline the fact that the definition of orbitally continuous in the setting of b-metric space is the same and it is valid.

Theorem 5.4 *We presume that a selfmapping T, defined on the b-metric space (X, d, s), is orbitally continuous. We also assume that the (X, d, s) is complete. If there is $\psi \in \Psi_b$ so that*

$$\min\{d(Tx, Ty), d(y, Ty), d(x, Tx)\} - \min d(Tx, y), d(x, Ty)\} \leq \psi(d(x, y)),$$
$$(5.11)$$

for each $x, y \in X$, then the sequence $\{T^n x_0\}_{n \in \mathbb{N}}$, for each $x_0 \in X$, converges to a fixed point of T.

Proof Set $x \in X$. We built up an iterative sequence $\{x_n\}$ as follows:

$$x_0 := x \text{ and } x_n = Tx_{n-1} \text{ for each } n \in \mathbb{N}. \tag{5.12}$$

We suppose that

$$x_n \neq x_{n-1} \text{ for each } n \in \mathbb{N}. \tag{5.13}$$

In particular, if for some $n_0 \in \mathbb{N}$ we have the inequality $x_{n_0} = Tx_{n_0-1} = x_{n_0-1}$, then, the proof is completed. By substituting $x = x_{n_0-1}$ and $y = x_{n_0}$ in (5.11), we find

$$\min\{d(Tx_{n-1}, Tx_n), d(x_{n-1}, Tx_{n-1}), d(x_n, Tx_n)\}$$
$$- \min\{d(x_{n-1}, Tx_n), d(Tx_{n-1}, x_n)\} \leq \psi(d(x_{n-1}, x_n)). \tag{5.14}$$

It implies that

$$\min\{d(x_n, x_{n+1}), d(x_n, x_{n-1})\} \leq \psi(d(x_{n-1}, x_n)). \tag{5.15}$$

Due to the fact that $\psi(t) < t$, for each $t > 0$, the case $d(x_n, x_{n-1}) \leq \psi(d(x_{n-1}, x_n))$ is impossible. Accordingly, we have

$$d(x_n, x_{n+1}) \leq \psi(d(x_{n-1}, x_n)). \tag{5.16}$$

Applying Remark 5.1 recurrently, we find that

$$d(x_n, x_{n+1}) \leq \psi(d(x_{n-1}, x_n)) \leq \psi^2(d(x_{n-2}, x_{n-1})) \leq \cdots \leq \psi^n(d(x_0, x_1)). \tag{5.17}$$

By Lemma 5.1, we deduce that

$$\lim_{n \to \infty} d(x_{n+1}, x_n) = 0. \tag{5.18}$$

Next we investigate whether the sequence $\{x_n\}$ is Cauchy.

Set $d(x_n, x_{n+k})$ for $k \geq 1$. Employing the extended triangle inequality, recursively, the following approximation is observed:

$$
\begin{aligned}
d(x_n, x_{n+k}) &\leq s[d(x_n, x_{n+1}) + d(x_{n+1}, x_{n+k})] \\
&\leq sd(x_n, x_{n+1}) + s\{s[d(x_{n+1}, x_{n+2}) + d(x_{n+2}, x_{n+k})]\} \\
&= sd(x_n, x_{n+1}) + s^2 d(x_{n+1}, x_{n+2}) + s^2 d(x_{n+2}, x_{n+k}) \\
&\ \ \vdots \\
&\leq sd(x_n, x_{n+1}) + s^2 d(x_{n+1}, x_{n+2}) + \cdots \\
&\quad + s^{k-1} d(x_{n+k-2}, x_{n+k-1}) + s^{k-1} d(x_{n+k-1}, x_{n+k}) \\
&\leq sd(x_n, x_{n+1}) + s^2 d(x_{n+1}, x_{n+2}) + \cdots \\
&\quad + s^{k-1} d(x_{n+k-2}, x_{n+k-1}) + s^k d(x_{n+k-1}, x_{n+k}),
\end{aligned}
\tag{5.19}
$$

since $s \geq 1$. Taking both (5.17) and (5.19) into consideration, we find

$$
\begin{aligned}
d(x_n, x_{n+k}) &\leq s\psi^n(d(x_0, x_1)) + s^2 \psi^{n+1} d(x_0, x_1) + \cdots \\
&\quad + s^{k-1} \psi^{n+k-2}(d(x_0, x_1)) + s^k \psi^{n+k-1}(d(x_0, x_1)) \\
&= \frac{1}{s^{n-1}} [s^n \psi^n(d(x_0, x_1)) + s^{n+1} \psi^{n+1} d(x_0, x_1) + \cdots \\
&\quad + s^{n+k-2} \psi^{n+k-2}(d(x_0, x_1)) + s^{n+k-1} \psi^{n+k-1}(d(x_0, x_1))].
\end{aligned}
\tag{5.20}
$$

Consequently, we have

$$
d(x_n, x_{n+k}) \leq \frac{1}{s^{n-1}} \left[P_{n+k-1} - P_{n-1} \right], \quad k \geq 1, n \geq 1,
\tag{5.21}
$$

with $P_n = \sum_{j=0}^{n} s^j \psi^j (d(x_0, x_1))$, $n \geq 1$. Consequently, by Lemma 5.1,

$\sum_{j=0}^{\infty} s^j \psi^j (d(x_0, x_1))$ is convergent. Upon taking limit $n \to \infty$ in (5.21) with keeping this observation in mind and regarding $s \geq 1$, we get

$$
\lim_{n \to \infty} d(x_n, x_{n+k}) \leq \lim_{n \to \infty} \frac{1}{s^{n-1}} \left[P_{n+k-1} - P_{n-1} \right] = 0.
\tag{5.22}
$$

We conclude that the sequence $\{x_n\}$ is Cauchy in (X, d).

Owing to the construction $x_n = T^n x_0$ and keeping on the mind that (X, p) is T-orbitally complete, there is $z \in X$ such that $x_n \to z$ and hence $x_n \to Tz$. Accordingly, $z = Tz$ which terminates the proof. □

Example

Let (X, d, s) be a b-metric space, where $X = \{1, 2, 3, 6\}$, $d : X \times X \to [0, \infty)$, where $d(x, y) = (x - y)^2$ and $s = 2$. Let $T : X \to X$, with

$$
T1 = T2 = 2, \ T3 = 1, \ T6 = 3.
$$

Of course, since $T^n x = x$, for $n \leq 3$ and any $x \in X$, the mapping T is orbitally continuous. Let

$$m_1(x, y) = \min\{d(Tx, Ty), d(y, Ty), d(x, Tx)\},$$
$$m_2(x, y) = \min\{d(Tx, y), d(x, Ty)\}.$$

Then, choosing $\psi \in \Psi_b$, $\psi(t) = \frac{t}{2}$, we get:

- For $x = 1$, $y = 2$,

$$m_1(1, 2) = 0, \ m_2(1, 2) = 0, \ d(1, 2) = 1,$$

  and (5.11) obviously holds.
- For $x = 1$, $y = 3$,

$$m_1(1, 3) = 1, \ m_2(1, 3) = 0, \ d(1, 3) = 4,$$
$$m_1(1, 3) - m_2(1, 3) = 1 < \frac{4}{2} = \psi(d(1, 3)).$$

- For $x = 1$, $y = 6$,

$$m_1(1, 6) = 1, \ m_2(1, 6) = 1, \ d(1, 6) = 25,$$
$$m_1(1, 6) - m_2(1, 6) = 0 < \frac{25}{2} = \psi(d(1, 6)).$$

- For $x = 2$, $y = 3$,

$$m_1(2, 3) = 0, \ m_2(2, 3) = 1, \ d(2, 3) = 1,$$
$$m_1(2, 3) - m_2(2, 3) = -1 < \frac{1}{2} = \psi(d(2, 3)).$$

- For $x = 2$, $y = 6$,

$$m_1(2, 6) = 0, \ m_2(2, 6) = 1, \ d(2, 6) = 16,$$
$$m_1(2, 6) - m_2(2, 6) = -1 < \frac{16}{2} = \psi(d(2, 6)).$$

- For $x = 3$, $y = 6$,

$$m_1(3, 6) = 4, \ m_2(3, 6) = 0, \ d(3, 6) = 9,$$
$$m_1(3, 6) - m_2(3, 6) = 4 < \frac{9}{2} = \psi(d(3, 6)).$$

Consequently, all the assumptions of Theorems 5.4 being verified, the mapping T has a fixed point, that is, $x = 2$.

The following corollary is obtained by letting $\psi(t) = kt$ where $k \in [0, 1)$.

Corollary 5.1 *We presume that a selfmapping T, defined on the b-metric space (X, d, s), is orbitally continuous. We also assume that the (X, d, s) is complete. If there is $k \in [0, 1)$ so that*

$$\min\{d(Tx, Ty), d(y, Ty), d(x, Tx)\} - \min\{d(Tx, y), d(x, Ty)\} \leq kd(x, y), \quad (5.23)$$

for each $x, y \in X$, then, the sequence $\{T^n x_0\}_{n \in \mathbb{N}}$ converges to a fixed point of T, for each $x_0 \in X$.

In case $s = 1$, the previous corollary coincides with the renowned Ćirić's non-unique fixed point theorem.

Corollary 5.2 (Non-unique fixed point theorem of Ćirić [97]) *We presume that a selfmapping T, defined on the b-metric space (X, d, s), is orbitally continuous. We also assume that the (X, d, s) is complete. If there is $k \in [0, 1)$ such that*

$$\min\{d(Tx, Ty), d(y, Ty), d(x, Tx)\} - \min\{d(Tx, y), d(x, Ty)\} \leq kd(x, y),$$

for each $x, y \in X$, then the sequence $\{T^n x_0\}_{n \in \mathbb{N}}$ converges to a fixed point of T, for each $x_0 \in X$.

Remark 5.2 Recalling the example that show that cone metric forms a b-metric, the analog of Ćirić non-unique fixed point theorem, Corollary 5.2, in the setting of cone metric space over a normal cone, is sustained (see e.g.[179]).

Theorem 5.5 *We presume that a selfmapping T, defined on the b-metric space (X, d, s), is orbitally continuous. We also assume that the (X, d, s) is complete and there are $a_1, a_2, a_3, a_4, a_5 \in \mathbb{R}_0^+$ such that*

$$0 \leq \frac{a_4 - a_2}{a_1 + a_2} < 1, \quad a_1 + a_2 \neq 0, \quad a_1 + a_2 + a_3 > 0 \text{ and } 0 \leq a_3 - a_5. \quad (5.24)$$

Define the function $K_T(x, y)$ as

$$K_T(x, y) = a_1 d(Tx, Ty) + a_2\big[d(x, Tx) + d(y, Ty)\big] + a_3[d(y, Tx) + d(x, Ty)]. \quad (5.25)$$

If the inequality

$$K_T(x, y) \leq a_4 d(x, y) + a_5 d(x, T^2 x) \quad (5.26)$$

holds for each $x, y \in X$, then, T possesses a fixed point.

Proof We skip the proof of this theorem since it can be obtained by a simple adaptation of the proof of Theorem 5.4.   □

Theorem 5.5 is still valid in the context of standard metric space.

Corollary 5.3 (See [156]) *We presume that a selfmapping T, defined on the standard metric space (X, d), is orbitally continuous. We also assume that the (X, d, s) is complete.*

Suppose there are $a_1, a_2, a_3, a_4, a_5 \in \mathbb{R}_0^+$ and a self mapping $T : X \to X$ satisfies the conditions

$$0 \leq \frac{a_4 - a_2}{a_1 + a_2} < 1, \quad a_1 + a_2 \neq 0, \ a_1 + a_2 + a_3 > 0 \text{ and } 0 \leq a_3 - a_5, \tag{5.27}$$

$$K_T(x, y) \leq a_4 d(x, y) + a_5 d(x, T^2 x) \tag{5.28}$$

hold for each $x, y \in X$, $K_T(x, y)$ is defined as in (5.25). Then, T has at least one fixed point.

Remark 5.3 Similarly to Remark 5.2, Theorem 5.5 can be characterized in the framework of cone metric spaces. In particular, the example (that indicates each cone metric with normal cone forms a b-metric) yields that Corollary 5.3 is sustained in the framework of cone metric space over normal cone (see e.g.[156]).

Theorem 5.6 *We presume that a selfmapping T, defined on the b-metric space (X, d, s), is orbitally continuous. We also assume that the (X, d, s) is complete and there is a (b)-comparison function $\psi \in \Psi_b$ such that*

$$\frac{P_T(x,y) - Q_T(x,y)}{R_T(x,y)} \leq \psi(d(x, y)), \tag{5.29}$$

for each $x, y \in X$, where

$$R_T(x, y) = \min\{d(y, Ty), d(x, Tx)\},$$
$$P_T(x, y) = \min\{d(x, Tx)d(y, Ty), d(Tx, Ty)d(x, y)\},$$
$$Q_T(x, y) = \min\{d(y, Ty)d(Tx, y), d(x, Tx)d(x, Ty)\}.$$

with $R_T(x, y) \neq 0$. Then, the sequence $\{T^n x_0\}_{n \in \mathbb{N}}$ converges to a fixed point of T, for each $x_0 \in X$.

Proof The proof is obtained by a simple and slide modification of the proof of Theorem 5.4.   □

Corollary 5.4 *We presume that a selfmapping T, defined on the b-metric space (X, d, s), is orbitally continuous. We also assume that the (X, d, s) is complete and there is $k \in [0, 1)$ such that*

$$\frac{P_T(x,y) - Q_T(x,y)}{R_T(x,y)} \leq kd(x, y), \tag{5.30}$$

for each $x, y \in X$, where

$$P_T(x, y) = \min\{d(Tx, Ty)d(x, y), d(x, Tx)d(y, Ty)\},$$
$$Q_T(x, y) = \min\{d(x, Tx)d(x, Ty), d(y, Ty)d(Tx, y)\},$$
$$R_T(x, y) = \min\{d(x, Tx), d(y, Ty)\}.$$

with $R_T(x, y) \neq 0$. Then, for each $x_0 \in X$ the sequence $\{T^n x_0\}_{n \in \mathbb{N}}$ converges to a fixed point of T.

Corollary 5.5 (Nonunique fixed point of Achari [4]) *We presume that a selfmapping T, defined on the standard metric space (X, d), is orbitally continuous. We also assume that the (X, d, s) is complete and there is $k \in [0, 1)$ so that*

$$\frac{P_T(x,y) - Q_T(x,y)}{R_T(x,y)} \leq kd(x, y), \tag{5.31}$$

for each $x, y \in X$, where

$$P_T(x, y) = \min\{d(x, Tx)d(y, Ty), d(Tx, Ty)d(x, y)\},$$
$$Q_T(x, y) = \min\{d(y, Ty)d(Tx, y), d(x, Tx)d(x, Ty)\},$$
$$R_T(x, y) = \min\{d(y, Ty), d(x, Tx)\}.$$

with $R_T(x, y) \neq 0$. Then, the sequence $\{T^n x_0\}_{n \in \mathbb{N}}$ converges to a fixed point of T, for each $x_0 \in X$.

Theorem 5.7 *We presume that a selfmapping T, defined on the b-metric space (X, d, s), is orbitally continuous. We also assume that the (X, d, s) is complete and there is $k \in [0, 1)$ such that*

$$m_T(x, y) - n_T(x, y) \leq kd(x, Tx)d(y, Ty), \tag{5.32}$$

for each $x, y \in X$, where

$$m_T(x, y) = \min\{[d(y, Ty)]^2, d(Tx, Ty)d(x, y), [d(Tx, Ty)]^2\},$$
$$n_T(x, y) = \min\{d(y, Ty)d(x, Tx), d(y, Tx)d(x, Ty)\}$$

with $R_T(x, y) \neq 0$. Then, the sequence $\{T^n x_0\}_{n \in \mathbb{N}}$ converges to a fixed point of T, for each $x_0 \in X$.

Proof The proof is observed easily by following the lines in the proof of Theorem 5.4 with the necessary motivation. □

Example

Let $X = \{\alpha, \beta, \gamma, \delta\}$, and the function $d : X \times X \to [0, \infty)$, with

$$d(x, y) = d(y, x), \ d(x, x) = 0, \ \text{for any } x, y \in X,$$
$$d(\alpha, \delta) = \tfrac{1}{2}, \ d(\alpha, \beta) = 1$$
$$d(\beta, \delta) = d(\gamma, \delta) = 4$$
$$d(\alpha, \gamma) = d(\beta, \gamma) = 2.$$

Thus, d forms a b-metric, with $s=3$.
Let also the mapping $T : X \times X$,

$$T\alpha = T\delta = \alpha, \ T\beta = T\gamma = \delta.$$

We consider the following cases:

- For $x = \alpha$, $y \in X$, we have $d(\alpha, T\alpha) = 0$ and then (5.32) holds.
- For $x = \beta$, $y = \gamma$, we have $m_T(\beta, \gamma) = 0$, (since $d(T\beta, T\gamma) = 0$), and (5.32) holds.
- For $x = \beta$, $y = \delta$,

$$m_T(\beta, \delta) = \tfrac{1}{4}, \ n_T(\beta, \delta) = 0, \ d(\beta, T\beta) = 4, \ d(\delta, T\delta) = \tfrac{1}{2},$$
$$m_T(\beta, \delta) - n_T(\beta, \delta) = \tfrac{1}{4} < k \cdot 2 = k \cdot d(\beta, T\beta)d(\delta, T\delta),$$

for any $k \in [0, 1)$.
- For $x = \gamma$, $y = \delta$,

$$m_T(\gamma, \delta) = \tfrac{1}{4}, \ n_T(\gamma, \delta) = 0, \ d(\gamma, T\gamma) = 4, \ d(\delta, T\delta) = \tfrac{1}{2},$$
$$m_T(\gamma, \delta) - n_T(\gamma, \delta)\tfrac{1}{4} < k \cdot 2 = k \cdot d(\gamma, T\gamma)d(\delta, T\delta),$$

for any $k \in [0, 1)$. Thus, since T is orbitally continuous, by Theorem (5.7) it follows then T has a fixed point, that is $x = \alpha$.

Theorem 5.5 is sustained in the framework of standard metric space.

Corollary 5.6 (Nonunique fixed point of Pachpatte [223]) *We presume that a selfmapping T, defined on the standard metric space (X, d), is orbitally continuous. We also assume that the (X, d, s) is complete. We define the following functions*

$$n_T(x, y) = \min\{d(y, Ty)d(x, Tx), d(y, Tx)d(x, Ty)\},$$
$$m_T(x, y) = \min\{[d(y, Ty)]^2, d(Tx, Ty)d(x, y), [d(Tx, Ty)]^2\}.$$

If there is $k \in [0, 1)$ such that

$$m_T(x, y) - n_T(x, y) \le kd(x, Tx)d(y, Ty), \tag{5.33}$$

for each $x, y \in X$ then, for each $x_0 \in X$ the sequence $\{T^n x_0\}_{n \in \mathbb{N}}$ converges to a fixed point of T.

Remark 5.4 An analog of Theorem 5.7 in the framework of cone metric spaces can be deduced as it mentioned in Remark 5.2.

Fixed Point Theorems in Partial Metric Spaces

In this chapter, we recollect some crucial and exciting fixed point theorems in the context of partial metric space. In addition, we underline the importance of the partial metric space in fixed point theory. Matthews [210] not only introduced the partial metric spaces but also obtained the first fixed point theorem in this new setting. More precisely, Matthews [210] showed that the famous Banach fixed point theory is valid in the framework of complete partial metric space. After this pioneering result of Matthews [210] in fixed point theory, a considerable number of researchers have investigated the partial metric spaces, and remarkable number of fixed point results in the context of partial metrics have appeared in the literature (see e.g. [1, 2, 7, 27–30, 34, 38–40, 42, 46, 49–53, 86, 92, 94, 102, 120, 125, 130, 136, 137, 146, 154, 155, 157, 164–166, 173, 175, 176, 178, 181, 188, 189, 202, 204, 210, 211, 222, 225, 230, 243, 245–250, 255, 258, 261, 262, 265, 266, 272–278] and the reference therein).

Although the notion of the partial metric is novel and has indispensable importance in the domain theory of the computer science, we should underline that not all fixed point results in partial metric spaces bring a novelty. More specifically, Haghi et al. [125] indicated that certain fixed point results in the setting of partial metric spaces are equivalent to relevant results in the context of the metric spaces. Indeed, for a mapping $T : X \to X$ where $X \neq \emptyset$ it was concluded that

$$M_d^T(x, y) = M_p^T(x, y), \text{ with}$$

$$M_\rho^T(x, y) = \max\{\rho(x, y), \rho(x, Tx), \rho(Ty, y), \rho(Tx, y), \rho(x, Ty)\}$$

where $\rho = d$, p are metric, partial metric, respectively. For this reason, we care about the presented results in this manuscript so that the techniques of Haghi et al. [125] are not applicable.

© The Author(s), under exclusive license to Springer Nature Switzerland AG 2022   97
E. Karapınar and R. P. Agarwal, *Fixed Point Theory in Generalized Metric Spaces*,
Synthesis Lectures on Mathematics & Statistics,
https://doi.org/10.1007/978-3-031-14969-6_6

6.1   Fixed Point Theorems via Admissible Mappings

Definition 6.1 A self-mapping T on a partial metric space (X, p) is called an $(\alpha - \psi)$-type K-contraction if there real numbers $0 \leq a_i \leq 1$, $i = 1, 2, 3, 4$, and $a_1 + a_2 + a_3 + 2a_4 \leq 1$, and if there are auxiliary functions $\psi \in \Psi$ and $\alpha : X \times X \to [0, \infty)$ so that

$$\alpha(x, y) p(Tx, Ty) \leq \psi(K(x, y)) \text{ for each } x, y \in X, \tag{6.1}$$

with

$$K(x, y) := a_1 p(x, y) + a_2 p(x, Tx) + a_3 p(y, Ty) + a_4 [p(y, Tx) p(x, Ty)]. \tag{6.2}$$

Theorem 6.1 *We presume that a selfmapping T on a complete partial metric space (X, p) forms α-orbital admissible and an $(\alpha - \psi)$-type K-contraction. If there is $x_0 \in X$ so that $\alpha(x_0, Tx_0) \geq 1$, then, there is $x^* \in X$ so that $p(x^*, x^*) = 0$ and $Tx^* = x^*$.*

Proof By assumption, $x_0 \in X$ with $\alpha(x_0, Tx_0) \geq 1$. We built up a recursive sequence $\{x_n\}$ in X by

$$x_{n+1} = Tx_n \text{ for each } n \in \mathbb{N}_0.$$

Observe that if $x_{n_0} = x_{n_0+1}$ for some n_0, then $x^* = x_{n_0}$ is a fixed point of T. Thus, throughout the proof, we presume that $x_n \neq x_{n+1}$ for each n. Regarding that T is α−admissible, we get

$$\alpha(x_0, x_1) = \alpha(x_0, Tx_0) \geq 1 \text{ implies } \alpha(Tx_0, Tx_1) = \alpha(x_1, x_2) \geq 1.$$

Inductively, we find

$$\alpha(x_n, x_{n+1}) \geq 1, \text{ for each } n \in \mathbb{N}_0. \tag{6.3}$$

Due to (6.1) and (6.3), it yields, for each $n \in \mathbb{N}$, that

$$p(x_{n+1}, x_n) = p(Tx_n, Tx_{n-1}) \leq \alpha(x_n, x_{n-1}) p(Tx_n, Tx_{n-1}) \leq \psi(K(x_n, x_{n-1})), \tag{6.4}$$

where

$$\begin{aligned}
K(x_n, x_{n-1}) &= a_1 p(x_n, x_{n-1}) + a_2 p(x_n, Tx_n) + a_3 p(x_{n-1}, Tx_{n-1}) \\
&\quad + a_4 [p(x_n, Tx_{n-1}) + p(x_{n-1}, Tx_n)] \\
&= a_1 p(x_n, x_{n-1}) + a_2 p(x_n, x_{n+1}) + a_3 p(x_{n-1}, x_n) \\
&\quad + a_4 [p(x_n, x_n) + p(x_{n-1}, x_{n+1})].
\end{aligned}$$

Taking (p_4) into account, we derive that

$$p(x_{n-1}, x_{n+1}) + p(x_n, x_n) \leq p(x_n, x_{n-1}) + p(x_n, x_{n+1})$$

Hence,

$$K(x_n, x_{n-1}) = (a_1 + a_3 + a_4) p(x_n, x_{n-1}) + (a_2 + a_4) p(x_n, x_{n+1}).$$

If for some $n \in \mathbb{N}$, we have $p(x_n, x_{n-1}) \leq p(x_n, x_{n+1})$, then we get

$$K(x_n, x_{n-1}) \leq (a_1 + a_2 + a_3 + 2a_4) p(x_n, x_{n+1}) \leq p(x_n, x_{n+1}), \qquad (6.5)$$

since $a_1 + a_2 + a_3 + a_4 \leq 1$. By taking (6.5) in consideration together with the fact that ψ is a nondecreasing function, we obtain from the inequality (6.4) that

$$p(x_{n+1}, x_n) \leq \psi(K(x_n, x_{n-1})) \leq \psi(p(x_n, x_{n+1})) < p(x_n, x_{n+1}),$$

a contradiction. Thus, for each $n \in \mathbb{N}$, we have

$$p(x_n, x_{n+1}) \leq p(x_n, x_{n-1}). \qquad (6.6)$$

Using (6.4) and (6.6), we get that

$$p(x_{n+1}, x_n) \leq \psi(p(x_n, x_{n-1})), \qquad (6.7)$$

for each $n \in \mathbb{N}$. By induction, we get

$$p(x_{n+1}, x_n) \leq \psi^n(p(x_1, x_0)), \text{ for each } n \in \mathbb{N}. \qquad (6.8)$$

Due to Lemma 4.1 (i), we find that

$$\lim_{n \to \infty} p(x_{n+1}, x_n) = 0. \qquad (6.9)$$

Again by keeping the expression (6.8) in the mind, and by using the triangular inequality (p_4), for each $k \geq 1$, we have

$$p(x_n, x_{n+k}) \leq p(x_n, x_{n+1}) + \ldots + p(x_{n+k-1}, x_{n+k}) - \sum_{j=1}^{k-1}(p(x_{n+j}, x_{n+j}))$$

$$\leq \sum_{j=n}^{n+k-1} \psi^j(p(x_1, x_0))$$

$$\leq \sum_{j=n}^{+\infty} \psi^j(p(x_1, x_0)) \to 0 \text{ as } n \to \infty.$$

This implies that

$$\lim_{n \to \infty} p(x_n, x_{n+k}) = 0,$$

and hence $\{x_n\}$ is a Cauchy sequence in (X, d). Since (X, p) is complete, there is $x^* \in X$ so that

$$\lim_{n \to \infty} p(x_n, x^*) = 0 = \lim_{n \to \infty} p(x_n, x_{n+k}) = p(x^*, x^*). \qquad (6.10)$$

Since T is continuous, by Definition 4.4, we conclude from (6.10) that

$$\lim_{n \to \infty} p(x_{n+1}, Tx^*) = \lim_{n \to \infty} p(Tx_n, Tx^*) = 0. \qquad (6.11)$$

On account of (6.10) and (6.11), Lemma 4.2 yields that x^* forms a fixed point of T. □

Example

Let $X = [0, \infty)$ and the partial metric $p : X \times X \to [0, \infty)$, where $p(x, y) = \max\{x, y\}$, for any $x, y \in X$. Let also, the mapping $T : X \to X$,

$$Tx = \begin{cases} 0, & \text{for } x \in [0, 1] \\ \left(\frac{1}{2}\right)^x, & \text{for } x \in (1, 2] \\ e^{1-2x}, & \text{for } x > 2. \end{cases}$$

Considering the function $\alpha : X \times X \to [0, \infty), \alpha(x, y) = \begin{cases} 1, & \text{for } x, y \in [0, 2] \\ 0, & \text{otherwise} \end{cases}$, it easy to see that T is an α-orbital admissible mapping. We choose the function $\psi \in \Psi$, defined by $\psi(t) = \frac{7}{8}t$ and the constants $a_1 = \frac{1}{2}, a_2 = a_3 = \frac{1}{6}, a_4 = 0$. Thus,

1. for $x, y \in [0, 1]$, we have $p(Tx, Ty) = 0$, and (6.1) holds;
2. for $x, y \in (1, 2]$, supposing, for example that $x > y$, we have

$$p(Tx, Ty) = \left(\frac{1}{2}\right)^y, \quad p(x, y) = x, \quad p(x, Tx) = x, \quad p(y, Ty) = y,$$

and then

$$\alpha(x, y)p(Tx, Ty) = \left(\frac{1}{2}\right)^y < \frac{1}{2} < \frac{7}{12} < \frac{7}{8}\left[\frac{1}{2}x + \frac{1}{6}(x + y)\right] = \psi(K(x, y));$$

3. for $x \in [0, 1], y \in (1, 2]$,

$$p(Tx, Ty) = \left(\frac{1}{2}\right)^y, \quad p(x, y) = y, \quad p(x, Tx) = x, \quad p(y, Ty) = y,$$

and

$$\alpha(x, y) p(Tx, Ty) = \left(\frac{1}{2}\right)^y < \frac{7}{12} < \frac{7}{8}\left[\frac{1}{2}y + \frac{1}{6}(x + y)\right] = \psi(K(x, y));$$

4. In the other cases, $\alpha(x, y) = 0$, and (6.1) holds.

Thereupon, T is an $(\alpha - \psi)$-type K-contraction, and from Theorem 6.1, there exists $x \in X$, $x = 0$, such that $T0 = 0$.

Theorem 6.2 *We presume that a selfmapping continuous T on a complete partial metric space (X, p) forms α-orbital admissible and an $(\alpha - \psi)$-type K-contraction. If (X, p) is regular and if there is $x_0 \in X$ so that $\alpha(x_0, Tx_0) \geq 1$, then, there is $x^* \in X$ so that $p(x^*, x^*) = 0$ and $Tx^* = x^*$.*

Proof It is clear that Theorem 6.1 differ from Theorem 6.2 each other at the requirement of the continuity of T and regularity of the space.

By verbatim of the related lines in the proof of Theorem 6.1, we have an iterative convergent sequence $\{x_n\}$ defined by $x_{n+1} = Tx_n$ for each $n \in \mathbb{N}_0$ with a limit $x^* \in X$. Taking the regularity of the space into account, (6.3) implies that there is a subsequence $\{x_{n(k)}\}$ of $\{x_n\}$ in a way that $\alpha(x_{n(k)}, x^*) \geq 1$ for each k. We derive, from (6.1), that

$$
\begin{aligned}
p(x_{n(k)+1}, Tx^*) = p(Tx_{n(k)}, Tx^*) &\leq \alpha(x_{n(k)}, x^*) p(Tx_{n(k)}, Tx^*) \\
&\leq \psi(K(x_{n(k)}x^*)) \\
&< K(x_{n(k)}, x^*x^*),
\end{aligned}
\tag{6.12}
$$

for each k, where,

$$
\begin{aligned}
K(x_{n(k)}, x^*) = &\, a_1 p(x_{n(k)}, x^*) + a_2 p(x_{n(k)}, x_{n(k)+1}) \\
&+ a_3 p(x^*, Tx^*) + a_4 [p(x_{n(k)}, Tx^*) + p(x^*, x_{n(k)+1})].
\end{aligned}
$$

Setting $k \to \infty$ in (6.12), we derive that

$$p(x^*, Tx^*) \leq (a_3 + a_4) p(x^*, Tx^*),
\tag{6.13}$$

a contradiction. Accordingly, we find $p(x^*, Tx^*) = 0$ and hence, $x^* = Tx^*$. □

Note that Theorems 6.1 and 6.2 guarantee only the existence of the fixed point. For the uniqueness of a fixed point, we consider the following hypothesis.

(U) For all $x^*, y^* \in Fix(T)$, then $\alpha(x^*, y^*) \geq 1$.

Theorem 6.3 *Putting condition to the statements of In addition to the hypotheses of Theorem 6.1 (resp. Theorem 6.2), if we assume the hypothesis (U) we deduce that the fixed point x^* of T is unique.*

Proof Suppose, on the contrary, that x^*, y^* are two distinct fixed point of T and hence $p(x^*, y^*) > 0$. Regarding that ψ is increasing and $\psi(0) = 0$ we find $\psi(p(x^*, y^*)) > 0$.

Taking (U) into consideration, the assumption of the Theorem 6.1 (resp. Theorem 6.2) yields

$$p(x^*, y^*) \leq \alpha(x^*, y^*)p(Tx^*, Ty^*)$$
$$\leq \psi(K(x^*, y^*)) = \psi(p(x^*, y^*))$$
$$< p(x^*, y^*),$$

a contradiction. Thus, $x^* = y^*$.   □

Definition 6.2 A self-mapping T on a partial metric space (X, p) is called an $(\alpha - \psi)$-type *N-contraction* if there real numbers $0 \leq a_i \leq 1$, $i = 1, 2, 3$, and $a_1 + 2a_2 + 2a_3 \leq 1$, and if there are auxiliary functions $\psi \in \Psi$ and $\alpha : X \times X \to [0, \infty)$ so that

$$\alpha(x, y)p(Tx, Ty) \leq \psi(N(x, y)) \text{ for each } x, y \in X, \tag{6.14}$$

where

$$N(x, y) := a_1 p(x, y) + a_2[p(x, Tx) + p(y, Ty)] + a_3[p(x, Ty) + p(y, Tx)]. \tag{6.15}$$

Theorem 6.4 *We presume that a selfmapping continuous T on a complete partial metric space (X, p) forms α-orbital admissible and an $(\alpha - \psi)$-type N-contraction. We also assume either (X, p) is regular or T is continuous. If there is $x_0 \in X$ so that $\alpha(x_0, Tx_0) \geq 1$, then, there is $x^* \in X$ so that $p(x^*, x^*) = 0$ and $Tx^* = x^*$.*

Keeping Theorems 6.1 and 6.2 in mind, the proof of Theorem 6.4 is obvious due the fact that $N(x, y) \leq K(x, y)$ for each $x, y \in X$.

Example

Let the set $X = [0, \infty)$ endowed with the partial metric $p : X \times X \to [0, \infty)$, $p(x, y) = \max\{x, y\}$ and $T : X \to X$, with

$$Tx = \begin{cases} \frac{x}{x^2+9}, & \text{if } x \in [0, 1] \\ \frac{1}{10}, & \text{if } x > 1 \end{cases}$$

be a continuous mapping on X. Choosing the function $\alpha : X \times X \to [0, \infty)$,

$$\alpha(x, y) = \begin{cases} 1, & \text{for } x, y \in [0, 1] \\ 2, & \text{for } x = 2, y = 0 \ , \\ 0, & \text{otherwise} \end{cases}$$

it follows easily that T is an α-orbital admissible mapping. We claim that, letting $a_1 = a_2 = \frac{1}{4}$, $a_3 = \frac{1}{16}$ and the function $\psi(t) = \frac{1}{3}t$, the mapping T forms an $(\alpha - \psi)$-type N-contraction.

Indeed the only interesting cases are:

1. $x, y \in [0, 1]$. We can assume that $x > y$. Then,

$$p(Tx, Ty) = \frac{x}{x^2+9}, \ p(x, y) = x, \ p(x, Tx) = x, \ p(y, Ty) = y,$$
$$p(x, Ty) = x, \ p(y, Tx) = \max \left\{ y, \frac{x}{x^2+9} \right\}$$
$$N(x, y) = \frac{1}{4}(x + x + y) + \frac{1}{16} \left(x + \max \left\{ y, \frac{x}{x^2+9} \right\} \right)$$

and

$$\alpha(x, y)p(Tx, Ty) = \frac{x}{x^2+9} \le \frac{3}{16}x < \frac{1}{3} \left[\frac{9}{16}x + \frac{1}{4}y + \frac{1}{16}\max \left\{ y, \frac{x}{x^2+9} \right\} \right] = \psi(N(x, y)).$$

2. $x = 2, y = 0$. We have

$$p(T2, T0) = \frac{1}{10}, \ p(2, 0) = 2, \ p(0, T0) = 0,$$
$$p(2, T2) = 2, \ p(0, T2) = \frac{1}{10}, \ p(2, T0) = 2$$
$$N(2, 0) = \frac{1}{4}[2 + 0 + \frac{1}{10}] + \frac{1}{16}[2 + \frac{1}{10}] = \frac{21}{32},$$

and then

$$\alpha(x, y)p(Tx, Ty) = 2 \cdot \frac{1}{10} < \frac{1}{3} \cdot \frac{21}{32} = \psi(N(2, 0)).$$

We conclude that the mapping T has a fixed point.

For the uniqueness condition for Theorem 6.4, the following hypothesis is considered:

(H) For all $x, y \in \text{Fix}(T)$, there is $z \in X$ so that $\alpha(x, z) \ge 1$ and $\alpha(y, z) \ge 1$,

where, $\text{Fix}(T)$ denotes the set of fixed points of T.

Theorem 6.5 *Adding condition to the hypotheses of In addition to the hypotheses of Theorem 6.4, we presume (H) to grantee that the fixed point x^* of T is unique.*

Proof Let y^* be a distinct fixed point of T. Due to (H), there is $z \in X$ so that

$$\alpha(x^*, z) \geq 1 \text{ and } \alpha(y^*, z) \geq 1. \tag{6.16}$$

Since x^*, y^* are the fixed point of T and T is α–admissible, the inequalities in (6.16) turns into

$$\alpha(x^*, T^n z) \geq 1 \text{ and } \alpha(y^*, T^n z) \geq 1, \text{ for each } n. \tag{6.17}$$

We construct an iterative sequence $\{z_n\}$ in X by $z_{n+1} = T z_n$ for each $n \in \mathbb{N}_0$ and $z_0 = z$. By (6.17), we have

$$p(x^*, z_{n+1}) = p(Tx^*, Tz_n) \leq \alpha(x^*, z_n)p(Tx^*, Tz_n) \leq \psi(K(x^*, z_n)), \tag{6.18}$$

for each n, where

$$\begin{aligned}
K(x^*, z_n) &= a_1 p(x^*, z_n) + a_2 p(x^*, Tx^*) + a_3 p(z_n, Tz_n) + a_3[p(x^*, Tz_n) + p(z_n, Tx^*)] \\
&= a_1 p(x^*, z_n) + a_2[p(x^*, x^*) + p(z_n, z_{n+1})] + a_3[p(x^*, z_{n+1}) + p(z_n, x^*)] \\
&\leq a_1 p(x^*, z_n) + a_2[p(x^*, z_{n+1}) + p(x^*, z_{n+1}) - p(x^*, x^*)] + a_4[p(x^*, z_{n+1}) + p(z_n, x^*)] \\
&= (a_1 + a_2 + a_2)p(x^*, z_n) + (a_2 + a_3 d)(x^*, z_{n+1}) + .
\end{aligned}$$

$\square$

We presume, without lost of the generality, that $p(x^*, z_n) > 0$ for each n. If we have $p(x^*, z_n) \leq p(x^*, z_{n+1})$, then due to the monotone property of ψ, and the inequality (6.18), we find

$$p(x^*, z_{n+1}) \leq \psi((a_1 + 2a_2 + 2a_3)p(x^*, z_{n+1})). \tag{6.19}$$

Recalling the monotone property of ψ together with (6.18), we suppose that $\max\{p(x^*, z_n), p(x^*, z_{n+1})\} = p(x^*, z_{n+1})$. Ergo, we find that, for each n,

$$p(x^*, z_{n+1}) \leq \psi((a_1 + 2a_2 + 2a_3)p(x^*, z_{n+1})) \leq \psi(p(x^*, z_{n+1})) < p(x^*, z_{n+1}),$$

a contradiction. Accordingly, we find $\max\{p(x^*, z_n), p(x^*, z_{n+1})\} = p(x^*, z_n)$, and

$$p(x^*, z_{n+1}) \leq \psi(p(x^*, z_n)),$$

for each n. This implies that

$$p(x^*, z_n) \leq \psi^n(p(x^*, z_0)), \text{ for each } n \in \mathbb{N}.$$

Setting $n \to \infty$ in the inequality above, we get

$$\lim_{n \to \infty} p(z_n, x^*) = 0. \tag{6.20}$$

Similarly, one can show that

$$\lim_{n \to \infty} p(z_n, y^*) = 0. \tag{6.21}$$

Taking (3.49) and (3.50) into account, we obtain $x^* = y^*$. As a result, x^* is the unique fixed point of T.

6.2   Nonunique Fixed Points Theorems

Let Y be a subset of a partial metric space (X, p) and T be a self-mapping on X. We give the characterization of basic notation for nonunique fixed point as follows we set (cf. [97])

1. $O(x, n) = \{x, Tx, T^2x, \dots, T^nx\}$ for $n \in \mathbb{N}$,
2. $O(x, \infty) = \{x, Tx, T^2x, \dots, \}$,
3. $\delta(Y) = \sup\{p(x, y) : x, y \in Y\}$.

Definition 6.3   *(cf.* [97])

1. A selfmapping T on a partial metric space (X, p) is called orbitally continuous if

$$\lim_{i,j\to\infty} p(T^{n_i}x, T^{n_j}x) = \lim_{i\to\infty} p(T^{n_i}x, z) = p(z, z) \qquad (6.22)$$

implies

$$\lim_{i,j\to\infty} p(TT^{n_i}x, TT^{n_j}x) = \lim_{i\to\infty} p(TT^{n_i}x, Tz) = p(Tz, Tz) \qquad (6.23)$$

for each $x \in X$.
2. A partial metric space (X, p) is called orbitally complete if each fundamental (Cauchy) sequence $\{T^{n_i}x\}_{i=1}^{\infty}$ converges in (X, p), that is, if

$$\lim_{i,j\to\infty} p(T^{n_i}x, T^{n_j}x) = \lim_{i\to\infty} p(T^{n_i}x, z) = p(z, z) \qquad (6.24)$$

Remark 6.1   Observe that for, any $m \in \mathbb{N}$, T^m is orbital continuous whenever T is orbital continuous.

Theorem 6.6   *We presume that a orbitally continuous selfmapping T, on a T-orbitally complete partial metric space (X, p), fulfills*

$$\min\{p(x, Tx), p(y, Ty), p(Tx, Ty)\} \le kp(x, y) \qquad (6.25)$$

for each $x, y \in X$ and for some $k \in (0, 1)$. Then, the $\{T^nx\}$ converges to a fixed point of T, for each $x \in X$.

Proof Starting from a random point $x_0 \in X$, we built-up a sequence

$$x_{n+1} = T x_n, \quad n \in \mathbb{N}_0. \tag{6.26}$$

If there is $n_0 \in \mathbb{N}$ so that $x_{n_0} = x_{n_0+1}$, then x_{n_0} forms a fixed point of T. This observation terminates the proof. Consequently, throughout the proof, we presume that $x_n \neq x_{n+1}$ for each $n \in \mathbb{N}_0$. By replacing $y = x_{n+1}$ and $x = x_n$ in (6.25) we observe that

$$\min\{p(x_n, T x_n), p(x_{n+1}, T x_{n+1}), p(T x_n, T x_{n+1})\} \leq kp(x_{n+1}, x_n)$$

which yields

$$\min\{p(x_n, x_{n+1}), p(x_{n+1}, x_{n+2})\} \leq kp(x_{n+1}, x_n). \tag{6.27}$$

Due to the hypothesis $k \in [0, 1)$, the inequality (6.27) gives $p(x_{n+1}, x_{n+2}) \leq kp(x_n, x_{n+1})$ for each $n = 0, 1, 2, \ldots$. Accordingly, we find

$$p(x_{n+1}, x_{n+2}) \leq kp(x_n, x_{n+1}) \leq k^2 p(x_{n-1}, x_n) \leq \cdots \leq k^{n+1} p(x_0, x_1). \tag{6.28}$$

We check whether the sequence $\{x_n\}$ is Cauchy. We presume, without loss of generality, that $n > m$. Accordingly, (6.28) and the weakened-triangle inequality (p_4) for partial metric implies that

$$0 \leq p(x_n, x_m) \leq p(x_n, x_{n-1}) + p(x_{n-1}, x_{n-2}) + \cdots + p(x_{m+1}, x_m)$$
$$-[p(x_{n-1}, x_{n-1}) + p(x_{n-2}, x_{n-2}) + \cdots p(x_{m+1}, x_{m+1})]$$

$$= \sum_{i=m}^{n} p(x_i, x_{i+1}) - \sum_{j=m+1}^{n-1} p(x_j, x_j)$$

$$\leq \sum_{i=m}^{n} p(x_i, x_{i+1})$$

$$\leq p(x_0, x_1) \sum_{s=m}^{n-1} k^s$$

$$= k^m \frac{1 - k^{n-m}}{1 - k} p(x_0, x_1).$$

Ergo, $\lim_{n,m \to \infty} p(x_n, x_m) = 0$ and hence the constructive sequence $\{x_n\}$ forms Cauchy in (X, p). On account of Lemma 4.1, the sequence $\{x_n\}$ also forms Cauchy in the corresponding metric space (X, d_p). Further, (X, p) is complete and its corresponding metric space (X, d_p) is complete. Consequently, regarding the corresponding complete metric space (X, d_p), there is $x^* \in X$ so that $x_n \to x^*$. Moreover, by Lemma 4.1

$$p(z, z) = \lim_{n \to \infty} p(x^*, x_n) = \lim_{n,m \to \infty} p(x_n, x_m) = 0,$$

which implies that

$$\lim_{n \to \infty} d_p(x^*, x_n) = 0. \tag{6.29}$$

Now, we check whether x^* is the fixed point of T. Due to (6.29), we have $p(x^*, x^*) = 0$. By replacing $y = x^*$ and $x = x_n$ in (6.25) we get

$$\min\{p(Tx_n, Tx^*), p(x_n, Tx_n), p(x^*, Tx^*)\} \le kp(x^*, x_n).$$

Accordingly, we find

$$\min\{p(x_{n+1}, Tx^*), p(x_n, x_{n+1}), p(x^*, Tx^*)\} \le kp(x^*, x_n). \tag{6.30}$$

Setting $n \to \infty$, we get

$$p(x^*, Tx^*) \le 0$$

using (6.29) and Lemma 4.5. Thus, $p(x^*, Tx^*) = 0$. Employing (4.4), we end up with

$$0 \le d_p(x^*, Tx^*) = 2p(x^*, Tx^*) - p(x^*, x^*) - p(Tx^*, Tx^*) = -p(Tx^*, Tx^*) \le 0.$$

As a result, $d_p(x^*, Tx^*) = 0$ and hence $x^* = Tx^*$, which completes the proof.   □

Example

Consider Matthews's basic partial metric example $p(x, y) = \max\{x, y\}$ on $X = \mathbb{R}^+$ so that (X, p) is a partial metric space (see e.g. [210, 211].) We presume that $T : X \to X$ is defined as $Tx = \frac{x^2}{1+3x}$ for each $x \in X$. We suppose, without loss of generality, that $x \ge y$. So, we have

$$p(Tx, Ty) = \max\left\{\frac{x^2}{1+3x}, \frac{y^2}{1+3y}\right\} = \frac{x^2}{1+3x}$$
$$p(Tx, x) = \max\left\{\frac{x^2}{1+3x}, x\right\} = x$$
$$p(y, Ty) = \max\left\{y, \frac{y^2}{1+3y}\right\} = y$$
$$p(x, y) = \max\{x, y\} = x$$

$$\min\{p(y, Ty), p(x, Tx), p(Tx, Ty)\} = \min\left\{\frac{x^2}{1+3x}, y, x\right\} = \min\left\{\frac{x^2}{1+3x}, y\right\} \tag{6.31}$$

For $k = \frac{1}{2}$, all hypotheses of Theorem 6.6. In particular, if $\min\{\frac{x^2}{1+3x}, y\} = \frac{x^2}{1+3x} \le \frac{x}{2}$. If $\min\{\frac{x^2}{1+3x}, y\} = y$ then $y \le \frac{x^2}{1+3x}$. Consequently, $y \le \frac{x^2}{1+3x} \le \frac{x}{2}$. We underline that $x = 0$ is the required fixed point of T.

Theorem 6.7 *Let $\varepsilon > 0$. We presume that a orbitally continuous selfmapping T on a T-orbitally complete partial metric space (X, p) satisfies*

$$0 < p(y, x) < \varepsilon \Rightarrow \min\{p(Ty, y), p(Tx, Ty), p(x, Tx)\} \leq kp(x, y) \qquad (6.32)$$

for each $x, y \in X$ and for some $k < 1$. If there is a point $x_0 \in X$ so that $p(x_0, T^n(x_0)) < \varepsilon$ for some $n \in \mathbb{N}$, then, T has a periodic point.

Proof Suppose $M = \{n \in \mathbb{N} : p(T^n x, x) < \varepsilon :$ for $x \in X\}$. Observe that $M \neq \emptyset$ due the statement of the theorem. Set $m = \min M$ so that $p(x, T^m x) < \varepsilon$ for $x \in X$. We examine two distinct cases: $m = 1$ or $m \geq 2$.

We presume that $m = 1$. So we have $p(Tx, x) < \varepsilon$. If we substitute $y = Tx$ in (6.32), we find

$$\min\{p(T(Tx), Tx), p(Tx, T(Tx)), p(x, Tx)\} \leq kp(Tx, x).$$

Since $k < 1$, the possibility $p(x, Tx) \leq kp(Tx, x)$ provides a contraction. Thus, $p(Tx, T(Tx)) = p(Tx, T^2 x) \leq kp(Tx, x)$. Verbatim of the initial lines in the proof of Theorem 6.6, we deduce that the iterative sequence $x_{n+1} = Tx_n$, $x = x_0$, converges to $x^* \in X$ so that $x^* = Tx^*$.

We presume that $m \geq 2$. Consequently, we find

$$p(Ty, y) \geq \varepsilon, \qquad (6.33)$$

for each $y \in X$. Accordingly, keeping (6.32) and $p(x, T^m x) < \varepsilon$ in mind, we derive

$$\min\{p(T(T^m x), T^m x), p(Tx, T(T^m x)), p(x, Tx)\} \leq kp(T^m x, x).$$

On account of $T^m x \in X$, we find $p(T(T^m x), T^m x) = p(T(w), w)$ where $T^m x = w$. Recalling (6.33), we get $p(T(T^m x), T^m x) = p(T(w), w) \geq \varepsilon$ and $p(Tx, x) \geq \varepsilon$. So,

$$\min\{p(T(T^m x), T^m x), p(Tx, T(T^m x)), p(x, Tx)\} = p(T^{m+1} x, Tx).$$

In particular,

$$p(T^{m+1} x, Tx) \leq kp(T^m x, x).$$

Recursively, one can get

$$p(T^2 x, T^{m+2} x) \leq p(Tx, T^{m+1} x) \leq k^2 p(x, T^m x).$$

Proceeding in this way, for each $s \in \mathbb{N}$, one can obtain

$$p(T^s x, T^{m+s} x) \leq p(T^{s-1} x, T^{m+s-1} x) \leq \cdots \leq k^s p(x, T^m x).$$

Thus, for the recursive sequence $x_{n+1} = T^m(x_n)$ where $x_0 = x$,

$$p(x_n, x_{n+1}) = p(T^{nm}(x_0), T^{(n+1)m}(x_0)) = p(T^{nm}(x_0), T^{m+nm}(x_0)) \leq k^{nm} p(x_0, T^m(x_0)).$$

By using the weakened-triangle inequality (p_4), for any $s \in \mathbb{N}$, one can get,

$$
\begin{aligned}
p(x_n, x_{n+s}) &\le \left[p(x_n, x_{n+1}) + p(x_{n+1}, x_{n+2}) + \cdots + p(x_{n+s-1}, x_{n+s}) \right] \\
&= k^{nm} \left[1 + k^m + \cdots + k^{(s-1)m} \right] p(x_0, T^m(x_0)) \\
&\le \frac{k^{nm}}{1-k^m} p(x_0, T^m(x_0)).
\end{aligned}
\tag{6.34}
$$

Thus,

$$
\lim_{n \to \infty} p(x_n, x_{n+s}) = 0.
$$

So $\{x_n\}$ is a Cauchy sequence in X. Taking T-orbitally completeness of X, there is some $x^* \in X$ so that

$$
\lim_{n \to \infty} p(T^{mn}(x_0), x^*) = \lim_{n \to \infty} p(x_n, x^*) = p(x^*, x^*) = 0.
$$

Regarding Remark 6.1, the orbital continuity of T implies that

$$
\begin{aligned}
p(T^m(x^*), T^m(x^*)) &= \lim_{n \to \infty} p(T^m(T^{nm}(x_0)), T^m x^*) = \lim_{n \to \infty} p(T^m(T^{nm}(x_0)), T^m(T^{nm}(x_0))) \\
&= \lim_{n \to \infty} (T^{(n+1)m}(x_0), T^m z) = \lim_{n \to \infty} (T^{(n+1)m}(x_0), T^{(n+1)m}(x_0)), \\
&= \lim_{n \to \infty} p(x_{n+1}, T^m x^*) = \lim_{n \to \infty} p(x_{n+1}, x_{n+1}) \\
&= p(x^*, T^m x^*) = p(x^*, x^*).
\end{aligned}
$$

So, we have $p(T^m x^*, T^m x^*) = p(x^*, T^m x^*) = p(x^*, x^*)$. Regarding (p_1), the point x^* is a periodic point of T. $\square$

Theorem 6.8 *We presume that a orbitally continuous selfmapping T on a T-orbitally complete partial metric space (X, p) satisfies*

$$
\min\{ p(Ty, y), p(Tx, Ty), p(Tx, x) \} < p(y, x)
\tag{6.35}
$$

for each $x, y \in X$, $x \ne y$. If the sequence $\{T^n(x_0)\}$ has a cluster point $z \in X$ for some $x_0 \in X$, then x^* is a fixed point of T.

Proof Suppose $T^m(x_0) = T^{m-1}(x_0)$ for some $m \in \mathbb{N}$, then $T^n(x_0) = T^m(x_0) = z$ for each $n \ge m$. It is clear that x^* is a required point.

Suppose $T^m(x_0) \ne T^{m-1}(x_0)$ for each $m \in \mathbb{N}$. Since $\{T^n(x_0)\}$ has a cluster point $z \in X$, one can write $\lim_{i \to \infty} T^{n_i}(x_0) = z$. Substituting y, x with $T^n(x_0), T^{n-1}(x_0)$, respectively, in (6.35),

$$
\begin{aligned}
&\min\{ p(T(T^n(x_0)), T^n(x_0)), p(T(T^{n-1}(x_0)), T(T^n(x_0))), p(T^{n-1}(x_0), T(T^{n-1}(x_0))) \} \\
&< p(T^n(x_0), T^{n-1}(x_0)).
\end{aligned}
\tag{6.36}
$$

The inequality $p(T^{n-1}(x_0), T^n(x_0)) < p(T^{n-1}(x_0), T^n(x_0))$ does not hold. Thus, (6.36) is equivalent to $p(T^n(x_0), T^{n+1}(x_0)) < p(T^{n-1}(x_0), T^n(x_0))$. Ergo, we have

$$\{p(T^n(x_0), T^{n+1}(x_0))\}_1^\infty \tag{6.37}$$

is decreasing and bounded below. Consequently $\{p(T^n(x_0), T^{n+1}(x_0))\}_1^\infty$ is convergent. On account of T-orbital continuity,

$$\lim_{i \to \infty} p(T^{n_i}(x_0), T^{n_i+1}(x_0)) = p(z, Tz). \tag{6.38}$$

Employing $\{p(T^{n_i}(x_0), T^{n_i+1}(x_0))\}_1^\infty \subset \{p(T^n(x_0), T^{n+1}(x_0))\}_1^\infty$ and (6.38), we have

$$\lim_{n \to \infty} p(T^n(x_0), T^{n+1}(x_0)) = p(z, Tz). \tag{6.39}$$

Considering the fact $\{p(T^{n_i+1}(x_0), T^{n_i+2}(x_0))\}_1^\infty \subset \{p(T^n(x_0), T^{n+1}(x_0))\}_1^\infty$ together with the limits $\lim_{i \to \infty} T^{n_i+1}(x_0) = Tz$, $\lim_{i \to \infty} T^{n_i+2}(x_0) = T^2z$ and (6.39) show that

$$p(Tz, T^2z) = p(z, Tz). \tag{6.40}$$

We presume $Tx^* \neq x^*$ and hence $p(x^*, Tx^*) > 0$. By replacing y, x with Tx^*, x^*, respectively, in (6.35) to get

$$\{p(x^*, T(x^*)), p(T(x^*), T(T(x^*))), p(T(T(x^*)), T(x^*))\} < p(x^*, T(x^*)). \tag{6.41}$$

which yields that $p(Tx^*, T^2x^*) < p(x^*, Tx^*)$. But, it contradicts (6.40). Consequently, we find $Tx^* = x^*$.

Theorem 6.9 *We presume that a orbitally continuous selfmapping T on a T-orbitally complete partial metric space (X, p) satisfies*

$$if \ 0 < p(x, y) < \varepsilon, \ then \ \min\{p(Ty, y), p(Tx, Ty), p(x, Tx)\} < p(x, y), \tag{6.42}$$

for each $x, y \in X$. If, for some $x_0 \in X$, the sequence $\{T^n(x_0)\}_{n=1}^\infty$ has a cluster point of $z \in X$, then x^* is a periodic point of T.

Proof Set $\lim_{i \to \infty} T^{n_i}(x_0) = x^*$, that is, for any $\varepsilon > 0$ there is $N_0 \in \mathbb{N}$ so that $p(T^{n_i}(x_0), x^*) < \frac{\varepsilon}{2K}$ for each $i > N_0$. Hence, by weakened-triangle inequality (p_4),

$$p(T^{n_i}(x_0), T^{n_i+1}(x_0)) \leq p(T^{n_i}(x_0), x^*) + p(x^*, T^{n_i+1}(x_0)) < \varepsilon.$$

Let us define the set

$$M = \{j \in \mathbb{N} : p(T^{n+j}(x_0), T^n(x_0)) < \varepsilon \ for \ some \ n \in \mathbb{N}\}.$$

We observe that $M \neq \emptyset$ due to the hypothesis of the theorem. Let $m = \min M$. The following options should be examined: $p(T^n(x_0), T^{n+m}(x_0)) = 0$ for some $n \in \mathbb{N}$ or

$p(T^n(x_0), T^{n+m}(x_0)) > 0$ for each $n \in \mathbb{N}$. First case yields $z = T^n(x_0) = T^{n+m}(x_0) = T^m(T^n(x_0)) = T^m(x^*)$. So, the claim of the theorem follows.

Now, we presume that $p(T^n(x_0), T^{n+m}(x_0)) > 0$ for each $n \in \mathbb{N}$. Set $r \in \mathbb{N}$ in a way that $p(T^r(x_0), T^{r+m}(x_0)) < \varepsilon$. If $m = 1$, and by substituting y, x with $T^{n+1}(x_0), T^n(x_0)$, respectively, in (6.42) we have

$$\min\{p(T(T^{n+1}(x_0)), T^{n+1}(x_0)), p(T(T^n(x_0)), T(T^{n+1}(x_0))), p(T^n(x_0), T(T^n(x_0)))\}$$
$$< p(T^{n+1}(x_0), T^n(x_0)). \tag{6.43}$$

Notice that the case $p(T^n(x_0), T^{n+1}(x_0)) < p(T^n(x_0), T^{n+1}(x_0))$ is impossible. Thus, (6.43) becomes

$$p(T^{n+1}(x_0), T^{n+2}(x_0)) < p(T^n(x_0), T^{n+1}(x_0)).$$

Ergo, $\{p(T^n(x_0), T^{n+1}(x_0))\}$ is decreasing for $n \geq r$. Consequently we find $Tx^* = x^*$ by a routine calculation.

Next, we presume $m \geq 2$. So,

$$p(T^n(x_0), T^{n+1}(x_0)) \geq \varepsilon, \tag{6.44}$$

for each $n \in \mathbb{N}$. So, $\lim_{i \to \infty} T^{n_i + r}(x_0) = T^r(x^*)$, since T is orbital continuous. Combining it with (6.44), we derive

$$p(T^r(x^*), T^{r+1}(x^*)) = \lim_{i \to \infty} p(T^{n_i + r}(x_0), T^{n_i + r + 1}(x_0)) \geq \varepsilon, \tag{6.45}$$

for each $r \in \mathbb{N}$. Recalling (6.42) together with the hypothesis $0 < p(T^j(x_0), T^{j+m}(x_0)) < \varepsilon$ we find,

$$\min\{p(T^j(x_0), T^{j+1}(x_0)), p(T^{j+1}(x_0), T^{j+m+1}(x_0)), p(T^{j+m}(x_0), T^{j+m+1}(x_0)).\}$$
$$< p(T^j(x_0), T^{j+m}(x_0)).$$

So, due to (6.44), we find $p(T^{j+1}(x_0), T^{j+m+1}(x_0)) < p(T^j(x_0), T^{j+m}(x_0)) < \varepsilon$. Continuing this process yields that

$$\cdots < p(T^{j+2}(x_0), T^{j+m+2}(x_0)) < p(T^{j+1}(x_0), T^{j+m+1}(x_0)) < p(T^j(x_0), T^{j+m}(x_0)) < \varepsilon. \tag{6.46}$$

As a result, $\{p(T^n(x_0), T^{n+m}(x_0)) : n \geq j\}$ is decreasing and bounded sequence. So, it is convergent. Further, the subsequence $\{p(T^{n_i}(x_0), T^{n_i+m}(x_0)) : i \in \mathbb{N}\}$ and $\{p(T^{n_i+1}(x_0), T^{n_i+1+m}(x_0)) : i \in \mathbb{N}\}$ are convergent to $d(z, T^m z)$ and $d(Tz, T^{m+1}z)$, respectively. By the orbital continuity of T and $\lim_{i \to \infty} T^{n_i}(x_0) = z$, one can get

$$p(T(z), T^{m+1}(z)) = p(z, T^m(z)) = \lim_{n \to \infty} p(T^n(x_0), T^{n+m}(x_0)). \tag{6.47}$$

So, we deduce $p(x^*, T^m x^*) < \varepsilon$ from (6.46) to (6.47). If $p(x^*, T^m x^*) = 0$, then $T^m x^* = x^*$ and the proof is completed. We presume that $p(x^*, T^m x^*) > 0$. Employing (6.42), we have

$$\min\{p(x^*, T(x^*)), p(T(x^*), T(T^m(x^*))), p(T(T^m(x^*)), T^m(x^*))\} < p(x^*, T^m x^*) < \varepsilon. \tag{6.48}$$

Taking (6.45), (6.48) into the account implies that $p(T(x^*), T^{m+1}(x^*)) < p(x^*, T^m x^*)$ which contradicts with (6.47). Ergo, $p(x^*, T^m x^*) = 0$, and so $T^m x^* = x^*$. $\qquad\square$

Theorem 6.10 *We presume that a orbitally continuous selfmapping T on a T-orbitally complete partial metric space (X, p) satisfies*

$$\min\{[[p(Ty, y)]^2, p(x, y)p(Tx, Ty), p(Tx, x)]^2\} \le kp(Ty, y)p(Tx, x), \tag{6.49}$$

for each $x, y \in X$ and for some $k < 1$. Then, $\{T^n x\}$ converges to a fixed point of T, for each $x \in X$,

Proof Verbatim of the corresponding initial lines in the proof of Theorem 6.8, we construct a recursive sequence $\{x_n\}$ by $x_{n+1} = Tx_n = T^{n+1}(x_0)$, for all $n \in \mathbb{N}$ by starting random $x_0 \in X$. This sequence forms a Cauchy incase of the existing some $n_0 \in \mathbb{N}$ with $x_{n_0+1} = x_{n_0}$. Keeping this observation in mind, in the rest of the proof, we presume that $x_{n+1} \ne x_n$ for each $n \in \mathbb{N}$. Substituting y, x with x_n, x_{n-1}, respectively, in (6.49), we find

$$\min\{[p(x_{n-1}, Tx_{n-1})]^2, p(x_{n-1}, x_n)p(Tx_{n-1}, Tx_n), [p(Tx_n, x_n)]^2\} \\ \le kp(x_{n-1}, Tx_{n-1})p(Tx_n, x_n). \tag{6.50}$$

Keeping $k < 1$ in mind, $p(x_n, x_{n+1})p(x_{n-1}, x_n) \le kp(x_n, x_{n+1})p(x_{n-1}, x_n)$ a contradiction. Accordingly, we have

$$p(x_n, x_{n+1}) \le kp(x_{n-1}, x_n).$$

Recursively, one can observe that

$$p(x_n, x_{n+1}) \le kp(x_{n-1}, x_n) \le k^2 p(x_{n-2}, x_{n-1}) \le \cdots \le k^n p(x_0, T(x_0)).$$

Verbatim of the related lines in the proof of Theorem 6.8, we deduce T possesses a fixed point. $\qquad\square$

Theorem 6.11 *Let T be a selfmapping on a nonempty set X endowed in two partial metrics ρ and p such that*

$$\min\{[\rho(y, Ty)]^2, \rho(x, y)\rho(Tx, Ty), [\rho(Tx, Ty)]^2\} \le k\rho(y, Ty)\rho(x, Tx) \tag{6.51}$$

for each $x, y \in X$ and for $0 \le k < 1$.

We presume that

(i)   T is orbitally continuous with respect to p,
(ii) $p(x, y) \leq \rho(x, y)$ for each $x, y \in X$,
(iii) X is orbitally complete space with respect to p.

Then, T possesses a fixed point in X.

Proof Verbatim of the corresponding initial lines in the proof of Theorem 6.8, we construct a recursive sequence $\{x_n\}$ by $x_{n+1} = T x_n = T^{n+1}(x_0)$, for all $n \in \mathbb{N}$ by starting random $x_0 \in X$.

Substituting y, x with x_n, x_{n-1}, respectively, in (6.51), one can get

$$\min\{[\rho(x_n, T x_n)]^2, \rho(x_{n-1}, x_n)\rho(T x_{n-1}, T x_n), [\rho(T x_{n-1}, T x_n)]^2\} \leq k\rho(x_n, T x_n)\rho(x_{n-1}, T x_{n-1}). \tag{6.52}$$

Because of the inequality $\rho(x_{n-1}, T x_{n-1})\rho(x_n, T x_n) \leq k\rho(x_{n-1}, T x_{n-1})\rho(x_n, T x_n)$, the expression in (6.52) is equivalent to $\rho(x_n, x_{n+1}) \leq k\rho(x_{n-1}, x_n)$. Recursively one can obtain

$$\rho(x_n, x_{n+1}) \leq k\rho(x_{n-1}, x_n) \leq \cdots \leq k^n \rho(x_0, x_1). \tag{6.53}$$

Recalling the weakened-triangle inequality (p_4), (6.53) implies that

$$\rho(x_n, x_{n+s}) \leq \frac{k^n}{1-k}\rho(x_0, x_1), \tag{6.54}$$

for any $s \in \mathbb{N}$. Due to the hypothesis (ii), we have

$$p(x_n, x_{n+p}) \leq \frac{k^n}{1-k}\rho(x_0, x_1). \tag{6.55}$$

Thus, $\{x_n\}$ is a Cauchy sequence with respect to p. So, there is $z \in X$ so that $\lim_{n \to \infty} T^n x = z$, since X is T-orbitally complete. From the orbital continuity of T, we find $Tz = \lim_{n \to \infty} T(T^n x) = z$. $\qquad \square$

Theorem 6.12 *We presume that a orbitally continuous selfmapping T on a T-orbitally complete partial metric space (X, p), for some $k \in (0, 1)$, satisfies*

$$\min\{p(T x, T y), p(x, T x), p(y, T y)\} - \min\{d_m^p(x, T y), d_m^p(T x, y)\} \leq k(p(x, y) - p(x, x)) + p(y, y), \tag{6.56}$$

for each $x, y \in X$. Then the sequence $\{T^n x_0\}_{n \in \mathbb{N}_0}$ converges with respect to τ_{d_p} to a fixed point of T, for each $x_0 \in X$.

Proof Take an arbitrary point $x_0 \in X$. We define the iterative sequence $\{x_n\}_{n \in \mathbb{N}_0}$ as follows:

$$x_{n+1} = Tx_n, \quad n \in \mathbb{N}_0.$$

If there is $n_0 \in \omega$ so that $x_{n_0} = x_{n_0+1}$, then x_{n_0} is a fixed point of T. Assume then that $x_n \neq x_{n+1}$ for each $n \in \mathbb{N}_0$.

Substituting $x = x_n$ and $y = x_{n+1}$ in (6.56) we find the inequality

$$\min\{p(x_{n+1}, x_{n+2}), p(x_n, x_{n+1}), p(x_{n+1}, x_{n+2})\}$$
$$- \min\{d_m^p(x_n, x_{n+2}), d_m^p(x_{n+1}, x_{n+1})\}$$
$$\leq k(p(x_n, x_{n+1}) - p(x_n, x_n)) + p(x_{n+1}, x_{n+1}).$$

By replacing $y = x_n$ and $x = x_{n+1}$ in (6.56), we find

$$\min\{p(x_{n+2}, x_{n+1}), p(x_{n+1}, x_{n+2}), p(x_n, x_{n+1})\}$$
$$- \min\{d_m^p(x_{n+1}, x_{n+1}), d_m^p(x_{n+2}, x_n)\}$$
$$\leq k(p(x_{n+1}, x_n) - p(x_{n+1}, x_{n+1})) + p(x_n, x_n),$$

which yields

$$\min\{p(x_n, x_{n+1}), p(x_{n+1}, x_{n+2})\}$$
$$\leq k(p(x_n, x_{n+1}) - p(x_n, x_n)) + p(x_{n+1}, x_{n+1}), \tag{6.57}$$

and

$$\min\{p(x_n, x_{n+1}), p(x_{n+1}, x_{n+2})\}$$
$$\leq k(p(x_n, x_{n+1}) - p(x_{n+1}, x_{n+1})) + p(x_n, x_n). \tag{6.58}$$

We presume that $p(x_{n_0}, x_{n_0+1}) \leq p(x_{n_0+1}, x_{n_0+2})$ for some $n_0 \in \omega$. Then, from the preceding two inequalities we conclude that

$$(1 - k)p(x_{n_0}, x_{n_0+1}) \leq \min\{p(x_{n_0+1}, x_{n_0+1}) - kp(x_{n_0}, x_{n_0}),$$
$$p(x_{n_0}, x_{n_0}) - kp(x_{n_0+1}, x_{n_0+1})\}.$$

If, for instance, $p(x_{n_0+1}, x_{n_0+1}) \leq p(x_{n_0}, x_{n_0})$, we have

$$(1 - k)p(x_{n_0}, x_{n_0+1}) \leq p(x_{n_0+1}, x_{n_0+1}) - kp(x_{n_0}, x_{n_0})$$
$$\leq (1 - k)p(x_{n_0+1}, x_{n_0+1})$$
$$\leq (1 - k)p(x_{n_0}, x_{n_0}),$$

so, by using $(P2)$, $p(x_{n_0}, x_{n_0+1}) = p(x_{n_0}, x_{n_0}) = p(x_{n_0+1}, x_{n_0+1})$, and hence $x_{n_0} = x_{n_0+1}$, a contradiction.

Hence, $p(x_n, x_{n+1}) > p(x_{n+1}, x_{n+2})$ for each $n \in \mathbb{N}_0$.

Respectively, by (6.57) we find

$$p(x_{n+1}, x_{n+2}) - p(x_{n+1}, x_{n+1}) \leq k(p(x_n, x_{n+1}) - p(x_n, x_n))$$
$$\leq k^2(p(x_{n-1}, x_n) - p(x_{n-1}, x_{n-1})) \qquad (6.59)$$
$$\leq \cdots \leq k^{n+1}((p(x_0, x_1) - p(x_0, x_0)),$$

for each $n \in \mathbb{N}_0$.

We indicate the sequence $\{x_n\}_{n \in \mathbb{N}}$ forms a Cauchy (X, p). Indeed, let $n, m \in \omega$ with $n < m$. Then, by using (6.59) and (p_4), we derive that

$$p(x_n, x_m) - p(x_n, x_n) \leq p(x_n, x_{n+1}) + \cdots + p(x_{m-1}, x_m) - \sum_{k=n}^{m-1} p(x_k, x_k)$$

$$\leq (k^n + \cdots + k^{m-1}) p(x_0, x_1).$$

So, the recursive sequence $\{x_n\}_{n \in \mathbb{2}}$ fulfills the hypothesis $(*)$ of Lemma 4.4. Ergo, it forms a Cauchy sequence in (X, p). Recalling the T-orbitally completeness of (X, p), there is $x^* \in X$ so that $x_n \to x^*$ with respect to τ_{d_p}. Further, we conclude $x_n \to Tx^*$ with respect to τ_{d_p} due to the orbital continuity of T. Ergo, $x^* = Tx^*$. □

Corollary 6.1 ([97, Theorem 1]) *If an orbitally continuous selfmapping T on T-orbitally complete metric space (X, d) satisfies*

$$\min\{d(Tx, Ty), d(y, Ty), d(x, Tx)\} - \min\{d(Tx, y), d(x, Ty)\}$$
$$\leq kd(x, y), \qquad (6.60)$$

for $k \in (0, 1)$ and for each $x, y \in X$, then $\{T^n x_0\}_{n \in \mathbb{N}_0}$ converges to a fixed point of T, for each $x_0 \in X$.

In what follows we give some examples so that Theorem 6.12 valid but Corollary 6.1 does not sustain for the metrics d_p and d_m^p, and p_0, respectively.

Example [176]

We define partial metric $p(x, y) = \max\{x, y\}$ on the set $X = \{0, 1, 2\}$. We construct the selfmapping T on X by $T0 = T1 = 0$ and $T2 = 1$.

Recalling that the metric space (X, p) is complete, we conclude that it is T-orbitally complete. In particular, T is orbitally continuous. A routine computation indicates

$$\min\{p(y, Ty), p(x, Tx), p(Tx, Ty)\} - \min\{d_m^p(Tx, y), d_m^p(x, Ty)\}$$
$$\leq \tfrac{1}{2}(p(x, y) - p(x, x)) + p(y, y),$$

for each $x, y \in X$. So, all hypotheses of Theorem 6.12 are attained. Consider now,

$$\min\{d_p(T1, T2), d_p(1, T1), d_p(2, T2)\} - \min\{d_p(1, T2), d_p(T1, 2)\}$$
$$= 1 - 0 = 1 > k = kd_p(1, 2),$$

for any $k \in (0, 1)$, so Corollary 6.1 does not sustain in the setting of complete metric space (X, d_p). In particular, it cannot be applied to (X, d_m^p), because $d_m^p = d_p$, in this case.

Example [176]

We define partial metric $p(x, y) = \max\{x, y\}$ on the set $X = [1, \infty)$. We construct the selfmapping T on X by $Tx = (x + 1)/2$ for each $x \in X$. We deduce T-orbitally complete, since (X, p) is complete. Note that T is is orbitally continuous, since it is continuous with respect to τ_{d_p}.

Now, we claim that T fulfils the hypothesis of (6.56) for any $k \in (0, 1)$. We examine two distinct possibilities for $x, y \in X$:

Case 1. $x = y$. So, we have

$$\min\{p(Tx, Ty), p(x, Tx), p(y, Ty)\} - \min\{d_m^p(x, Ty), d_m^p(Tx, y)\}$$
$$= \min\left\{\frac{x+1}{2}, x, x\right\} - \left(x - \frac{x+1}{2}\right) = 1$$
$$\leq x = p(x, x) = k((p(x, y) - p(x, x)) + p(y, y).$$

Case 2. $x \neq y$. We presume, without loss of generality, that $x > y$.
If $Tx \geq y$, we get

$$\min\{p(y, Ty), p(x, Tx), p(Tx, Ty)\} - \min\{d_m^p(Tx, y), d_m^p(x, Ty)\}$$
$$= \min\{\frac{x+1}{2}, x, y\} - \min\{x - \frac{y+1}{2}, \frac{x+1}{2} - y\}$$
$$= y - (\frac{x+1}{2} - y) = 2y - \frac{x+1}{2}$$
$$\leq y = p(y, y) = k((p(x, y) - p(x, x)) + p(y, y).$$

If $Tx < y$, we obtain

$$\min\{p(y, Ty), p(x, Tx), p(Tx, Ty)\} - \min\{d_m^p(Tx, y), d_m^p(x, Ty)\}$$
$$= \min\{\frac{x+1}{2}, x, y\} - \min\{x - \frac{y+1}{2}, y - \frac{x+1}{2}\}$$
$$= \frac{x+1}{2} - (y - \frac{x+1}{2}) = x + 1 - y$$
$$< y = p(y, y) = k((p(x, y) - p(x, x)) + p(y, y).$$

So, all hypotheses of Theorem 6.12 are fulfilled. In particular, T possesses a (unique) fixed point, namely, $x = 1$.

Observe that Corollary 6.1 can not sustain for T in the complete metric space (X, p_0). More precisely, for a given $k \in (0, 1)$, select $x > 1$ so that $x + 1 > 2kx$, and set $y = Tx$. Ergo,

$$\min\{p_0(Tx, Ty), p_0(x, Tx), p_0(y, Ty)\} - \min\{p_0(x, Ty), p_0(Tx, y)\}$$
$$= \min\{\frac{x+1}{2}, x\} - \min\{x, 0\} = \frac{x+1}{2} > kx = kp_0(x, y).$$

Consequently, (6.60) is not fulfilled.

The next results is the natural extension of [97, Theorem 3] in the framework of partial metric spaces.

Theorem 6.13 *We presume an orbitally continuous selfmapping T on a a partial metric space (X, p) fulfills*

$$\min\{p(Tx, Ty), p(x, Tx), p(y, Ty)\} - \min\{d_m^p(x, Ty), d_m^p(Tx, y)\} \tag{6.61}$$
$$< p(x, y) - p(x, x) + p(y, y),$$

for each $x, y \in X$ with $x \neq y$. If for some $x_0 \in X$ the sequence $\{T^n x_0\}_{n \in \mathbb{2}}$ has a cluster point $z \in X$ with respect to τ_{d_p}, then x^* is a fixed point of T.

Proof We presume $x_0 \in X$ in a way that the sequence $\{T^n x_0\}_{n \in \mathbb{2}}$ has a cluster point $z \in X$ with respect to τ_{d_p}. We construct a recursive sequence $\{x_n\}_{n \in \mathbb{N}_0}$ as $x_{n+1} = Tx_n, \quad n \in \mathbb{N}_0$.

If there is $n_0 \in \omega$ so that $x_{n_0} = x_{n_0+1}$, then x_{n_0} is a fixed point of T. Assume then that $x_n \neq x_{n+1}$ for each $n \in \mathbb{N}_0$.

Verbatim of the corresponding lines in the proof of Theorem 6.12, replacing $y = x_{n+1}$ and $x = x_n$ in (6.61) we derive

$$\min\{p(x_n, x_{n+1}), p(x_{n+1}, x_{n+2})\} < p(x_n, x_{n+1}) - p(x_n, x_n) + p(x_{n+1}, x_{n+1}),$$

and substituting $y = x_n$ and $x = x_{n+1}$ in (6.61), we get

$$\min\{p(x_n, x_{n+1}), p(x_{n+1}, x_{n+2})\} < p(x_n, x_{n+1}) - p(x_{n+1}, x_{n+1}) + p(x_n, x_n).$$

If $p(x_{n_0}, x_{n_0+1}) \leq p(x_{n_0+1}, x_{n_0+2})$ for some $n_0 \in \omega$, we deduce from the preceding two inequalities that $p(x_{n_0}, x_{n_0}) < p(x_{n_0+1}, x_{n_0+1})$ and $p(x_{n_0+1}, x_{n_0+1}) < p(x_{n_0}, x_{n_0})$, respectively, a contradiction.

So, $p(x_n, x_{n+1}) > p(x_{n+1}, x_{n+2})$ for each $n \in \mathbb{N}_0$, and thus the sequence $\{p(T^n x_0, T^{n+1} x_0)\}_{n \in \mathbb{N}_0}$ is convergent. Since $\{T^n x_0\}_{n \in \mathbb{N}_0}$ has a cluster point $z \in X$ with respect to τ_{d_p}, then there is a subsequence $\{T^{n_i} x_0\}_{i \in \omega}$ of $\{T^n x_0\}_{n \in \mathbb{N}_0}$ which converges to z with respect to τ_{d_p}. By the orbital continuity of T we have $T^{n_i+1} x_0 \to T x^*$ with respect to τ_{d_p}, so by Lemma 4.2,

$$\lim_{i \to \infty} p(T^{n_i} x_0, T^{n_i+1} x_0) = p(x^*, T x^*). \tag{6.62}$$

Therefore

$$\lim_{n \to \infty} p(T^n x_0, T^{n+1} x_0) = p(x^*, T x^*). \tag{6.63}$$

Again, by the orbital continuity of T we have $T^{n_i+2} x_0 \to T^2 z$ with respect to τ_{d_p} and hence

$$\lim_{n \to \infty} p(T^{n+1} x_0, T^{n+2} x_0) = p(T x^*, T^2 x^*),$$

so

$$p(T x^*, T^2 x^*) = p(x^*, T x^*). \tag{6.64}$$

Assume $T x^* \neq x^*$, that is, $p(x^*, T x^*) > 0$. So, one can replace x and y with x^* and $T x^*$, respectively, in (6.61) to deduce that

$$\min\{p(x^*, T x^*), p(T x^*, T^2 x^*)\} < p(x^*, T x^*),$$

which yields that $p(T x^*, T^2 x^*) < p(x^*, T x^*)$. This contradicts the equality (6.64). Thus, $T x^* = x^*$. The proof is complete.

Motivated by Ćirić's theorems [97], Pachpatte proved in [223, Theorem 1] that if T is an orbitally continuous self-map of a T-orbitally complete metric space (X, d) so that there is $k \in (0, 1)$ with

$$\min\{[d(Tx, Tx)]^2, d(x, y)d(Tx, Ty), [d(Ty, y)]^2\}$$
$$- \min\{d(x, Tx)d(y, Ty), d(x, Ty)d(y, Tx)\} \le kd(x, Tx)d(Ty, y) \tag{6.65}$$

for each $x, y \in X$, then for each $x_0 \in X$ the sequence $\{T^n x_0\}_{n \in \mathbb{N}_0}$ converges to a fixed point of T.

However, Pachpatte's theorem has a very limited field of application since under its conditions, if we denote by x^* any fixed point of T, it follows that for each $y \in X$, $Ty = x^*$ or $Ty = y$. Indeed, let $y \neq x^*$ and suppose $Ty \neq x^*$. Then from

$$\min\{[d(Tx^*, Ty)]^2, d(x^*, y)d(Tx^*, Ty), [d(y, Ty)]^2\}$$
$$- \min\{d(x^*, Tx^*)d(y, Ty), d(x^*, Ty)d(y, Tx^*)\}$$
$$\le kd(x^*, Tx^*)d(y, Ty),$$

it follows

$$\min\{[d(x^*, Ty)]^2, d(x^*, y)d(x^*, Ty), [d(y, Ty)]^2\} = 0.$$

Hence $d(y, Ty) = 0$, i.e., $y = Ty$.

In our next result we modify the contraction condition (6.65) and thus obtain a new fixed point theorem that avoids the inconvenient indicated above. In fact, this will be done in the more general setting of partial metric spaces and, to this end, the following notation will be used: If p is a partial metric on a set X we denote by p' the function defined on $X \times X$ by $p'(x, y) = p(x, y) - p(x, x)$ for each $x, y \in X$. (Of course, $p' = p$ whenever p is a metric on X.)

Theorem 6.14 *We presume that a orbitally continuous selfmapping T on a T-orbitally complete partial metric space (X, p), for some $k \in (0, 1)$, satisfies*

$$\min\{[[p'(y, Ty)]^2, p'(x, y)p'(Tx, Ty), p'(x, Tx)]^2\}$$
$$- \min\{d_m^p(x, Ty)d_m^p(y, Tx), d_m^p(x, Tx)d_m^p(y, Ty)\} \qquad (6.66)$$
$$\leq k \min\{[p'(x, y)]^2, p'(x, Tx)p'(y, Ty)\},$$

for each $x, y \in X$, then for each $x_0 \in X$ the sequence $\{T^n x_0\}_{n \in \mathbb{N}_0}$ converges with respect to τ_{d_p} to a fixed point of T.

Proof Verbatim of the relates lines of the proof of Theorem 6.12, for an arbitrary point $x_0 \in X$, we built-up an iterative sequence $\{x_n\}_{n \in \mathbb{N}_0}$ as $Tx_n = x_{n+1}$, $n \in \mathbb{N}_0$.

If there is $n_0 \in \omega$ so that $x_{n_0+1} = x_{n_0}$, then x_{n_0} forms a fixed point of T and hence it completes the proof. For this reason, for the rest of the proof, we presume that $x_n \neq x_{n+1}$ for each $n \in \mathbb{N}_0$.

By replacing $y = x_{n+1}$ and $x = x_n$ in (6.66), the following estimation is observed:

$$\min\{[p'(x_{n+1}, x_{n+2})]^2, p'(x_n, x_{n+1})p'(x_{n+1}, x_{n+2}), [p'(x_n, x_{n+1})]^2\}$$
$$\leq k \min\{[p'(x_n, x_{n+1})]^2, p'(x_n, x_{n+1})p'(x_{n+1}, x_{n+2})\}. \qquad (6.67)$$

The expression (6.67) implies that

$$\min\{[p'(x_{n+1}, x_{n+2})]^2, p'(x_n, x_{n+1})p'(x_{n+1}, x_{n+2}), [p'(x_n, x_{n+1})]^2\}$$
$$= [p'(x_{n+1}, x_{n+2})]^2,$$

and consequently, we have

$$p'(x_{n+1}, x_{n+2}) \leq kp'(x_n, x_{n+1}),$$

for each $n \in \mathbb{N}_0$. As a result, we find

$$p(x_n, x_{n+1}) - p(x_n, x_n) \leq k^n(p(x_0, x_1) - p(x_0, x_0)),$$

for each $n \in \mathbb{N}$. As in the proof of Theorem 6.12, we find $\{x_n\}_{n \in \mathbb{N}_0}$ forms a Cauchy sequence in (X, p). Recalling that (X, p) is T-orbitally complete and since $x_n = T^n x_0$ for each n, there is $z \in X$ so that $x_n \to z$ with respect to τ_{d_p}. Due to the orbital continuity of T, we get $x_n \to Tx^*$ with respect to τ_{d_p}. So, $x^* = Tx^*$ which completes the proof.

Corollary 6.2 *We presume that a orbitally continuous selfmapping T on a T-orbitally complete partial metric space (X, p), for some $k \in (0, 1)$, satisfies*

$$
\begin{aligned}
\min\{[[d(y, Ty)]^2, d(x, y)d(Tx, Ty), d(x, Tx)]^2\} \\
- \min\{d(x, Ty)d(y, Tx), d(x, Tx)d(y, Ty)\} \\
\leq k \min\{d(x, Tx)d(y, Ty), [d(x, y)]^2\},
\end{aligned} \tag{6.68}
$$

for each $x, y \in X$. Then, $\{T^n x_0\}_{n \in \mathbb{N}_0}$ converges to a fixed point of T, for each $x_0 \in X$.

Remark 6.2 Consider Matthews's basic partial metric example $p(x, y) = \max\{x, y\}$ on $X = \mathbb{R}^+$. On this structure any orbitally continuous selfmapping T on X with $Tx \leq x$ (for each $x \in X$) possesses a fixed point. In particular, for such a T we have $p'(x, Tx) = 0$ for each $x \in X$, so hypothesis (6.66) in Theorem 6.14, is fulfilled in a straightforward way.

In what follows, we indicate that Theorem 6.14 possesses a fixed point although Corollary 6.2 does not yield a result for any of induced metrics d_p, d_m^p and p_0.

Example [see [176]]

We consider Matthews's basic partial metric example $p(x, y) = \max\{x, y\}$ on $X = \mathbb{R}^+$. For this structure, we construct a selfmapping T as follows:

$$
Tx := \begin{cases} x - 1 & \text{if } x \geq 2, \\ 0 & \text{otherwise.} \end{cases}
$$

We observe that the selfmapping T forms orbitally continuous since for each $x \in X$ one has $T^n x \to 0$ with respect to τ_{d_p}, and $T0 = 0$. In addition, by Remark 6.2 the contraction condition (6.66) is also fulfilled. Accordingly, all the hypotheses of Theorem 6.14 hold.

Next, consider $y = Tx$ and $x \geq 3$. It yields $y \geq 2$ and $x - y = 1$. Consequently,

$$
\begin{aligned}
\min\{[d_p(y, Ty)]^2, d_p(x, y)d_p(Tx, Ty), [d_p(x, Tx)]^2\} \\
- \min\{d_p(x, Ty)d_p(y, Tx), d_p(x, Tx)d_p(y, Ty)\} \\
= \min\{1, (x - y)^2, 1\} - 0 = 1 \\
= \min\{d_p(x, Tx)d_p(y, Ty), [d_p(x, y)]^2\}.
\end{aligned}
$$

Ergo, (6.68) is not fulfilled for any $k \in (0, 1)$. Accordingly, we cannot apply Corollary 6.2 to (X, d_p), and hence to (X, d_m^p) and the selfmapping T.

Consequently, for $k \in (0, 1)$, select $x \geq 3$ so that $x > 1/(1 - k)$, with $y = Tx$. So,

$$\min\{[p_0(x, Tx)]^2, p_0(x, y)p_0(Tx, Ty), [p_0(y, Ty)]^2\}$$
$$- \min\{p_0(x, Tx)p_0(y, Ty), p_0(x, Ty)p_0(y, Tx)\}$$
$$= \min\{x^2, x(x - 1), (x - 1)^2\} - 0 = (x - 1)^2$$
$$> kx(x - 1)$$
$$= k\min\{p_0(x, Tx)p_0(y, Ty), [p_0(x, y)]^2\}.$$

So, we cannot apply Corollary 6.2 to (X, p_0) and the selfmapping T, although, T is orbitally continuous for (X, p_0).

References

1. Abdeljawad, T., Karapınar, E., Taş, K.: A generalized contraction principle with control functions on partial metric spaces. Comput. Math. Appl. **63**, 716–719 (2012)
2. Abedelljawad, T., Karapınar, E., Taş, K.: Existence and uniqueness of common fixed point on partial metric spaces. Appl. Math. Lett. **24**, 1894–1899 (2011)
3. Achari, J.: Results on non-unique fixed points. Publications de L'Institut Mathématique **26**, 5–9 (1978)
4. Achari, J.: On Ćirić's non-unique fixed points. Mat. Vesnik **13**, 255–257 (1976)
5. Afshari, H., Aydi, H., Karapınar, E.: Existence of Fixed Points of Set-Valued Mappings in b-Metric Spaces. East Asian Math. J. **32**, 319–332 (2016)
6. Agarwal, R.P., Karapınar, E., O'Regan, D., Roldan-Lopez-de-Hierro, A.F.: Fixed Point Theory in Metric Type Spaces. Springer International Publishing (2015)
7. Agarwal, R.P., Alghamdi, M.A., Shahzad, N.: Fixed point theory for cyclic generalized contractions in partial metric spaces. Fixed Point Theory Appl. **2012**, 2012.40. (2012)
8. Agarwal, R.P., Alsulami, H., Karapınar, E., Khojasteh, F.: Remarks on some recent fixed point results in quaternion-valued metric spaces. Abstr. Appl. Anal. **2014**, Article ID: 171624 (2014)
9. Aksoy, U., Karapınar, E., Erhan, I.M.: Fixed points of generalized α-admissible contractions on b-metric spaces with an application to boundary value problems. J. Nonlinear Convex **A17**, 1095–1108 (2016)
10. Alber, Y.I., Guerre-Delabriere, S.: Principle of weakly contractive maps in Hilbert spaces. In: Gohberg, I., Lybich, Y. (eds.) New Results in Operator Theory and its Applications, Operator Theory: Advances and Applications, vol. 98, pp. 7–22. Birkhäuser, Basel, Switzerland (1997)
11. Algahtani, B., Fulga, A., Karapınar, E.: Non-unique fixed point results in extended b-metric space. Mathematics **6**(5), 68 (2018)
12. Alghamdi, M.A., Hussain, N., Salimi, P.: Fixed point and coupled fixed point theorems on b-metric-like spaces. J. Inequal. Appl. **2013**, 402 (2013)
13. Alghamdi, M.A., Karapınar, E.: $G - \beta - \psi$ Contractive type mappings and related fixed point theorems. J. Inequal. Appl. **2013**, Article ID: 70, (2013)
14. Alghamdi, M.A., Karapınar, E.: $G - \beta - \psi$ Contractive type mappings in G-metric spaces. Fixed Point Theory Appl. **2013**, Article ID: 123 (2013)

© The Editor(s) (if applicable) and The Author(s), under exclusive license to Springer
Nature Switzerland AG 2022
E. Karapınar and R. P. Agarwal, *Fixed Point Theory in Generalized Metric Spaces*,
Synthesis Lectures on Mathematics & Statistics,
https://doi.org/10.1007/978-3-031-14969-6

15. Alharbi, A.S., Alsulami, H., Karapınar, E.: On the power of simulation and admissible functions in metric fixed point theory. J. Funct. Spaces **2017**, 2068163 (2017)

16. Ali, M.U., Kamram, T., Karapınar, E.: An approach to existence of fixed points of generalized contractive multivalued mappings of integral type via admissible mapping. Abstr. Appl. Anal. **2014**, Article ID 141489 (2014)

17. Ali, M.U., Kamran, T., Karapınar, E.: On (α, ψ, η)-contractive multivalued mappings. Fixed Point Theory Appl. **2014**, 7 (2014)

18. Ali, M.U., Kamram, T., Karapınar, E.: An approach to existence of fixed points of generalized contractive multivalued mappings of integral type via admissible mapping. Abstr. Appl. Anal. **2014**, Article ID: 141489 (2014)

19. AlMezel, S., Chen, C.-M., Karapınar, E., Rakočević, V.: Fixed point results for various α-admissible contractive mappings on metric-like spaces. Abstr. Appl. Anal. **2014**, Article ID: 379358 (2014)

20. Alqahtani, B., Fulga, A., Karapınar, E.: Sehgal type contractions on b-metric space. Symmetry, **2018**, 10, 560 (2018) https://doi.org/10.3390/sym10110560

21. Alharbi, A.S., Alsulami, H.H., Karapınar, E.: On the power of simulation and admissible functions in metric fixed point theory. J. Funct. Spaces **2017**, Article ID 2068163 (2017)

22. AlSulami, H., Gulyaz, S., Karapınar, E., Erhan, I.M.: Fixed point theorems for a class of alpha-admissible contractions and applications to boundary value problem. Abstr. Appl. Anal. **2014**, Article ID: 187031 (2014)

23. Alsulami, H.H., Karapınar, E., Rakočević, V.: Ciric type nonunique fixed point theorems on b-metric spaces. Filomat. **31**, 3147–3156 (2017)

24. Alsulami, H.H., Karapınar, E., Khojasteh, F., Roldán-López-de-Hierro, A.F.: A proposal to the study of contractions in quasi-metric spaces. Discret. Dyn. Nat. Soc., **2014**, Article ID 269286 (2014)

25. Alsulami, H., Almezel, S., Karapınar, E., Khojasteh, F.: A note on fixed point results in complex valued metric spaces. J. Inequal. Appl. **2015**, 2015:33 (2015)

26. Alsulami, H.H., Karapınar, E., Agarwal, R.P., Khojasteh, F.: A short note on C*-valued contraction mappings. J. Inequal. Appl. **2016**, 50 (2016)

27. Altun, I., Simsek, H.: Some fixed point theorems on dualistic partial metric spaces. J. Adv. Math. Stud. **1**, 1–8 (2008)

28. Altun, I., Sola, F., Simsek, H.: Generalized contractions on partial metric spaces. Topol. Appl. **157**, 2778–2785 (2010)

29. Altun, I., Acar, O.: Fixed point theorems for weak contractions in the sense of Berinde on partial metric spaces. Topol. Appl. **159**, 2642–2648 (2012)

30. Altun I., Erduran, A.: Fixed point theorems for monotone mappings on partial metric spaces. Fixed Point Theory Appl. **2011**, Article ID 508730 (2011)

31. Argoubi, H., Samet, B., Vetro, C.: Nonlinear contractions involving simulation functions in a metric space with a partial order. J. Nonlinear Sci. Appl. **8**, 1082–1094 (2015)

32. Arshad, M., Ameer, E., Karapınar, E.. Generalized contractions with triangular α-orbital admissible mapping on Branciari metric spaces. J. Inequal. Appl. **2016**, 2016:63 (2016)

33. An, T.V., Tuyen, L.Q., Dung, N.V.: Stone-type theorem on b -metric spaces and applications. Topol. Appl. **185**, 50–64 (2015)

34. Aydi, H., Abbas, M., Vetro, V.: Partial Hausdorff metric and Nadler's fixed point theorem on partial metric spaces. Topol. Appl. **159**, 3234–3242 (2012)

35. Aydi, H., Bilgili, N., Karapınar, E.: Common fixed point results from quasi-metric spaces to G-metric spaces. J. Egyptian Math. Soc. **23**(2), 356–361 (2015)

36. Aydi, H., Bota M., Karapınar, E., Mitrović, S.: A fixed point theorem for set-valued quasi-contractions in b-metric spaces. Fixed Point Theory Appl. **2012**, Article ID 88 (2012)

37. Aydi, H., Bota, M., Karapınar, E., Moradi, S.: A common fixed point for weak ϕ-contractions in b-metric spaces. Fixed Point Theory **13**, 337–346 (2012)
38. Aydi, H.: Common fixed point results for mappings satisfying (ψ, ϕ)-weak contractions in ordered partial metric space. Int. J. Math. Stat. **12**, 53–64 (2012)
39. Aydi, H.: Fixed point results for weakly contractive mappings in ordered partial metric spaces. J. Adv. Math. Stud. **4**, 1–12 (2011)
40. Aydi, H.: Fixed point theorems for generalized weakly contractive condition in ordered partial metric spaces. J. Nonlinear Anal. Optim.: Theory Appl. **2**, 33–48 (2011)
41. Aydi, H., Jellali, M., Karapınar, E.: On fixed point results for α-implicit contractions in quasi-metric spaces and consequences. Nonlinear Anal. Model. Control. **21**, 40–56 (2016)
42. Aydi, H., Jellali, M., Karapınar, E.: Common fixed points for generalized α-implicit contractions in partial metric spaces: Consequences and application. RACSAM - Revista de la Real Academia de Ciencias Exactas, Fisicas y Naturales. Serie A. Matematicas **109**, 367–384 (2015)
43. Aydi, H., Karapınar, E., Yazidi, H.: Modified T-contractions via α-admissible mappings and application to integral equations. Filomat. **31**(5), 1141–1148 (2017)
44. Aydi, H., Karapınar, E., Zhang, D.: A note on generalized admissible-Meir-Keeler-contractions in the context of generalized metric spaces. Results Math. **71**, 73–92 (2017)
45. Aydi, H., Karapınar, E., Rakočević, V.: Nonunique fixed point theorems on b-metric spaces via simulation functions. Jordan J. Math. Stat **12**(3), 265–288 (2019)
46. Aydi, H., Karapınar, E., Shatanawi, W.: Coupled fixed point results for (ϕ, φ)-weakly contractive condition in ordered partial metric spaces. Comput. Math. Appl. **62**(12), 4449–4460 (2011)
47. Aydi, H., Karapınar, E., Samet, B.: Fixed points for generalized (α, ψ)-contractions on generalized metric spaces. J. Inequal. Appl. **2014** 2014:229 (2014)
48. Aydi, H., Karapınar, E., Lakzian, H.: Fixed point results on the class of generalized metric spaces. Math. Sci. **6** (2012)
49. Aydi, H., Karapınar, E.: A Meir-Keeler common type fixed point theorem on partial metric spaces. Fixed Point Theory Appl. **2012**, 2012:26 (2012)
50. Aydi, H.: Some fixed point results in ordered partial metric spaces. J. Nonlinear Sci. Appl. **4**(2), 210–217 (2011)
51. Aydi, H.: Some coupled fixed point results on partial metric spaces. Int. J. Math. Math. Sci. **2011**, Article ID 647091 (2011)
52. Aydi, H., Vetro, C., Karapınar, E.: On Ekeland's variational principle in partial metric spaces. Appl. Math. Inf. Sci. **9**, 257–262 (2015)
53. Aydi, H., Vetro, C., Sintunavarat, W., Kumam, P.: Coincidence and fixed points for contractions and cyclical contractions in partial metric spaces. Fixed Point Theory Appl. **2012**, 2012:124 (2012)
54. Azam, A., Arshad, M.: Kannan fixed point theorems on generalized metric spaces. J. Nonlinear Sci. Appl. **1**, 45–48 (2008)
55. Bailey, D.F.: Some theorems on contractive mappings. J. Lond. Math. Soc. **41**, 101–106 (1966)
56. Banach, S.: Sur les opérations dans les ensembles abstraits et leur application aux équations intégrales. Fundam. Math. **3**, 133–181 (1922)
57. Bakhtin, I.A.: The contraction mapping principle in quasimetric spaces. Funct. Anal. Unianowsk Gos. Ped. Inst. **30**, 26–37 (1989)
58. Berinde, V.: Generalized contractions in quasi-metric spaces. Seminar on Fixed Point Theory, Babeş-Bolyai University, Research Sem., pp. 3–9 (1993)
59. Berinde, V.: Sequences of operators and fixed points in quasimetric spaces. Mathematica **41**, 23–27 (1996)
60. Berinde, V.: Contracţii Generalizate şi Aplicaţii, vol. 2. Editura Cub Press, Baie Mare, Romania (1997)

61. Berinde, V.: Iterative Approximation of Fixed Points. Editura Efemeride, Baia Mare (2002)
62. Berinde, V.: Common fixed points of noncommuting almost contractions in cone metric spaces. Math. Commun. **15**(1), 229–241 (2010)
63. Berinde, V.: Common fixed points of noncommuting discontinuous weakly contractive mappings in cone metric spaces. Taiwan. J. Math. **14**(5), 1763–1776 (2010)
64. Berinde, V.: Some remarks on a fixed point theorem for Ćirić-type almost contractions. Carpathian J. Math. **25**(2), 157–162 (2009)
65. Bessaga, C.: On the converse of the Banach "fixed-point principle". Colloq. Math. **7**, 41–43 (1959)
66. Bianchini, R.M., Grandolfi, M.: Transformazioni di tipo contracttivo generalizzato in uno spazio metrico. Atti Acad. Naz. Lincei, VII. Ser., Rend., Cl. Sci. Fis. Mat. Natur. **45**, 212–216 (1968)
67. Bilgili, N., Karapınar E., Samet, B.: Generalized $\alpha - \psi$ contractive mappings in quasi-metric spaces and related fixed-point theorems. J. Inequal. Appl. **2014**, 2014:36 (2014)
68. Bilgili, N., Karapınar, E.: A note on common fixed points for (ψ, α, β)-weakly contractive mappings in generalized metric spaces. Fixed Point Theory Appl. **2013**, 2013:287 (2013)
69. Bohl, P.: über die Bewegung eines mechanischen Systems in der Nähe einer Gleichgewichtslage. J. für dieReine und Angewandte Mathematik **127**, 179–276 (1904)
70. Boriceanu, M.: Strict fixed point theorems for multivalued operators in b-metric spaces. Int. J. Mod. Math. **4**, 285–301 (2009)
71. Boriceanu, M.: Fixed point theory for multivalued generalized contraction on a set with two b-metrics. Mathematica **54**, 3–14 (2009)
72. Boriceanu, M., Petruşel, A., Rus, I.A.: Fixed point theorems for some multivalued generalized contractions in b-metric spaces. Int. J. Math. Stat. **6**, 65–76 (2010)
73. Bourbaki, N.: Topologie Generale. Paris, France, Herman (1974)
74. Bota, M.: Dynamical Aspects in the Theory of Multivalued Operators. Cluj University Press (2010)
75. Bota, M., Molnár, A., Varga, C.: On Ekeland's variational principle in b-metric spaces. Fixed Point Theory **12**, 21–28 (2012)
76. Bota, M.F., Karapınar, E.: A note on "Some results on multi-valued weakly Jungck mappings in b-metric space.". Cent. Eur. J. Math. **11**, 1711–1712 (2013). https://doi.org/10.2478/s11533-013-0272-2
77. Bota, M., Karapınar E., Mleşniţe, O.: Ulam-Hyers stability for fixed point problems via $\alpha - \phi$-contractive mapping in b-metric spaces. Abstr. Appl. Anal. **2013** Article Id 855293 (2013)
78. Bota, M., Chifu, C., Karapınar, E.: Fixed point theorems for generalized $(\alpha - \psi)$-Ciric-type contractive multivalued operators in b-metric spaces. J. Nonlinear Sci. Appl. **9**, 1165–1177 (2016)
79. Bota, M.: Fixed point theorems for operators with a contractive iterate in b-metric spaces. Stud. Univ. Babes-Bolyai Math. **61**, 435–442 (2016)
80. Boyd, D.W., Wong, J.S.W.: On nonlinear contractions. Proc. Am. Math. Soc. **20**, 458–464 (1969)
81. Branciari, A.: A fixed point theorem of Banach-Caccioppoli type on a class of generalized metric spaces. Publ. Math. Debrecen **57**, 31–37 (2000)
82. Branciari, A.: A fixed point theorem for mappings satisfying a general contractive condition of integral type. Int. J. Math. Math. Sci. **29**, 531–536 (2002)
83. Brouwer, L.E.J.: Uber Abbildung von Mannigfaltigkeiten. Math. Ann. **71**, 97–115 (1912)
84. Browder, F.E.: On the convergence of successive approximations for nonlinear functional equations. Indag. Math. **30**, 27–35 (1968)
85. Bryant, V.W.: A remark on a fixed point theorem for iterated mappings. Am. Math. Mon. **75**, 399–400 (1968)

86. Bukatin, M., Kopperman, R., Matthews, S., Pajoohesh, H.: Partial metric spaces. Am. Math. Monthly **116**, 708–718 (2009)
87. Burton, T.A.: Integral equations, implicit functions and fixed points. Proc. Am. Math. Soc. **124**, 2383–2390 (1996). https://doi.org/10.1090/S0002-9939-96-03533-2
88. Caccioppoli, R.: Una teorema generale sull'esistenza di elementi uniti in una transformazione funzionale. Ren. Accad. Naz Lincei **11**, 794–799 (1930)
89. Chatterjea, S.K.: Fixed-point theorems. C. R. Acad. Bulg. Sci. **25**, 727–730 (1972)
90. Caristi, J.: Fixed point theorems for mappings satisfying inwardness conditions. Trans. Am. Math. Soc. **215**, 241–251 (1976)
91. Chen, C.M., Karapınar E., Lin, I.J.: Periodic points of weaker Meir-Keeler contractive mappings on generalized quasi-metric spaces. Abstr. Appl. Anal. **2014**, Article ID: 490450 (2014)
92. Chen, C.M., Karapınar, E.: Fixed point results for the α-Meir-Keeler contraction on partial Hausdorff metric spaces. J. Inequal. Appl. **2013**, 410 (2013)
93. Chen, C.M., Abkar, A., Ghods, S., Karapınar, E.: Fixed point theory for the α-admissible Meir-Keeler type set contractions having KKM* property on almost convex sets. Appl. Math. Inf. Sci. **11**(1), 171–176 (2017)
94. Chi, K.P., Karapınar, E., Thanh, T.D.: A generalized contraction principle in partial metric spaces. Math. Comput. Model. **55**, 1673–1681 (2012). https://doi.org/10.1016/j.mcm.2011.11.005
95. Chu, S.C., Diaz, J.B.: Remarks on a generalization of Banach's principle of contraction mappings. J. Math. Anal. Appl. **11**, 440–446 (1965)
96. Ćirić, L.: On Sehgal's maps with a contractive iterate at a point. Publ. De L'Institut Mathématique. **33**(47), 59–62 (1983)
97. Ćirić, L.B.: On some maps with a nonunique fixed point. Publ. Inst. Math. **17**, 52–58 (1974)
98. Ćirić, L.B.: Fixed Point Theory: Contraction Mapping Principle. C-print, Beograd (2003)
99. Ćirić, Lj.B.: Fixed points for generalized multi-valued mappings. Mat. Vesnik. **9** (24), 265–272 (1972)
100. Ćirić, L.B.: A generalization of Banach's contraction principle. Proc. Am. Math. Soc. **45**(2), 267–273 (1974)
101. Ćirić, L.B., Jotić, N.: A further extension of maps with non-unique fixed points. Mat. Vesnik **50**, 1–4 (1998)
102. Ćirić, L.j., Samet, B., Aydi, H., Vetro, C.: Common fixed points of generalized contractions on partial metric spaces and an application. Appl. Math. Comput. **218**, 2398–2406 (2011)
103. Czerwik, S.: Contraction mappings in b-metric spaces. Acta Math. et Inf. Uni. Ostraviensis. **1**, 5–11 (1993)
104. Dass, B.K., Gupta, S.: An extension of Banach contraction principle through rational expression. Indian J. Pure Appl. Math. **6**, 1455–1458 (1975)
105. Das, P., Lahiri, B.K.: Fixed point of a Ljubomir Ćirić's quasi-contraction mapping in a generalized metric space. Publ. Math. Debrecen **61**, 589–594 (2002)
106. Ding, H.S., Li, L.: Coupled fixed point theorems in partially ordered cone metric spaces. Filomat. **25**(2), 137–149 (2011)
107. Došenovic, T., Postolache M., Radenovic, S.:On multiplicative metric spaces: survey. Fixed Point Theory Appl. **2016**, 92 (2016)
108. Du, W.S.: A note on cone metric fixed point theory and its equivalence. Nonlinear Anal. **72**(5), 2259–2261 (2010)
109. Dunford, N., Schwartz, J.T.: Linear Operators. Part I, General Theory. Interscience, New York (1958)
110. Dutta, P.N., Choudhury, B.S.: A generalisation of contraction principle in metric spaces. Fixed Point Theory Appl. **2008**, 406368 (2008). https://doi.org/10.1155/2008/406368

111. Edelstein, M.: An extension of Banach's contraction principle. Proc. Am. Math. Soc. **12**, 7–10 (1961)
112. Edelstein, M.: On fixed and periodic points under contractive mappings. J. Lond. Math. Soc. **37**, 74–79 (1962)
113. Erhan, I.M., Karapınar E., Sekulic, T.: Fixed points of (psi,phi) contractions on rectangular metric spaces. Fixed Point Theory Appl. **2012** 2012:138 (2012)
114. Fisher, B.: On a theorem of Khan. Riv. Math. Univ Parma. **44**(1978), 135–137 (1978)
115. Fulga A., E. Karapınar E.: Revisiting Meir-Keeler type fixed operators on Branciari distance space. Tbilisi Math. J. **12**(4), 97–110 (2019)
116. Fréchet, M.R.: Sur quelques points du calcul fonctionnel. Rend. Circ. Mat. Palermo **22**, 1–74 (1906). https://doi.org/10.1007/BF03018603
117. George, R., Radenovic, S., Reshma, K.P., Shukla, S.: Rectangular b-metric space and contraction principles. J. Nonlinear Sci. Appl. **8**, 1005–1013 (2015)
118. Geraghty, M.: On contractive mappings. Proc. Am. Math. Soc. **40**, 604–608 (1973)
119. Gulyaz, S., Karapınar, E., Erhan, I.M.: Generalized α-Meir-Keeler contraction mappings on Branciari b-metric Spaces. Filomat **31**(17), 5445–5456 (2017)
120. Gulyaz, S., Karapınar, E.: Coupled fixed point result in partially ordered partial metric spaces through implicit function. Hacet. J. Math. Stat. **42**, 347–357 (2013)
121. Gupta S., Ram, B.: Non-unique fixed point theorems of Ćirić type. (Hindi) Vijnana Parishad Anusandhan Patrika **41**, 217–231 (1998)
122. Guseman, Jr.L.F.: Fixed point theorems for mappings with a contractive iterate at a point, Proc. Amer. Math. Soc. **26**, 615–618 (1970)
123. Hadamard, J.: Note sur quelques applications de l'indice de Kronecker in Jules Tannery: Introduction á la théorie des fonctions d'une variable (Volume 2), 2nd edn, vol. 1910, pp. 437–477. A. Hermann & Fils, Paris (1910)
124. Hadžić, O., Pap, E.: A fixed point theorem for multivalued mappings in probabilistic metric spaces and an application in fuzzy metric spaces. Fuzzy Sets Syst. **127**, 333–344 (2002)
125. Haghi, R.H., Rezapour, S., Shahzad, N.: Be careful on partial metric fixed point results. Topol. Appl. **160**(3), 450–454 (2013)
126. Haghi, R.H., Rezapour, S., Shahzad, N.: Some fixed point generalizations are not real generalizations. Nonlinear Anal. Theory Methods Appl. **74**(5), 1799–1803 (2011)
127. Hammache, K., Karapınar, E., Ould-Hammouda, A.: On Admissible weak contractions in b-metric-like space. J. Math. Anal. **8**(3), 167–180 (2017)
128. Hardy, G.E., Rogers, T.D.: A generalization of a fixed point theorem of Reich. Canad. Math. Bull. **16**, 201–206 (1973)
129. Hausdorff, F.: Mengenle hre. W. de Gruyter & Co (1927)
130. Heckmann, R.: Approximation of metric spaces by partial metric spaces. Appl. Categ. Struct. **7**, 71–83 (1999)
131. Hicks, T.L.: Fixed point theory in probabilistic metric spaces. Univ. u Novom Sadu Zb. Rad. Prirod.-Mat. Fak. Ser. Mat. **13**, 63–72 (1983)
132. Hicks, T.L.: Fixed point theorems for quasi-metric spaces. Math. Japonica. **33**, 231–236 (1988)
133. Hitzler, P., Seda, A.: Mathematical Aspects of Logic Programming Semantics, Studies in Informatics Series. Chapman and Hall, CRC Press, Taylor and Francis Group (2011)
134. Huang, L.G., Zhang, X.: Cone metric spaces and fixed point theorems of contractive mappings. J. Math. Anal. Appl. **332**(2), 1468–1476 (2007)
135. Hussain, N., Kadelburg, Z., Radenović, S., Al-Solamy, F.: Comparison functions and fixed point results in partial metric spaces. Abstr. Appl. Anal. **2012**, Article ID 605781 (2012)
136. Ilić, D., Pavlović, V., Rakočević, V.: Some new extensions of Banach's contraction principle to partial metric space. Appl. Math. Lett. **24**, 1326–1330 (2011)

137. Ilić, D., Pavlović, V., Rakočević, V.: Extensions of the Zamfirescu theorem to partial metric spaces. Original Research Article Math. Comput. Modell. **55**, 801–809 (2012)
138. Jachymski, J.: Fixed points of maps with a contractive iterate at a point. Math. Balkanica, New Ser. **9** (4) (1995)
139. Jachymski, J.: Equivalent conditions for generalized contractions on (ordered) metric spaces. Nonlinear Anal. **74**(3), 768–774 (2011)
140. Jachymski, J., Jozwik, I.: Nonlinear contractive conditions: a comparison and related problems, fixed point theory and its applications. Polish Acad. Sci. **77**, 123–146 (2007)
141. Jachymski, J.: Equivalence of some contractivity properties over metrical structures. Proc. Am. Math. Soc. **125**(8), 2327–2335 (1997)
142. Jaggi, D.S.: Some unique fixed point theorems. Indian J. Pure Appl. Math. **8**, 223–230 (1977)
143. Janković, S., Kadelburg, Z., Radenović, S.: On cone metric spaces: a survey. Nonlinear Anal. **74**, 2591–2601 (2011)
144. Janoş, L.: A converse of Banach's contraction theorem. Proc. Am. Math. Soc. **18**, 287–289 (1967)
145. Jleli, M., Karapınar, E., Samet, B.: Best proximity points for generalized $\alpha - \psi$-proximal contractive type mappings. J. Appl. Math. **2013**, Article ID 534127 (2013)
146. Jleli, M., Karapınar, E., Samet, B.: Further remarks on fixed point theorems in the context of partial metric spaces. Abstr. Appl. Anal. **2013** Article Id: 715456 (2013)
147. Jleli, M., Karapınar , E., Samet, B.: Further generalizations of the Banach contraction principle. J. Inequal. Appl. **2014**, 2014:439 (2014)
148. Jleli, M., Karapınar, E., Samet, B.: Fixed point results for $\alpha - \psi_\lambda$ contractions on gauge spaces and applications. Abstr. Appl. Anal. **2013**, Article ID 730825 (2013)
149. Jleli, M., Samet, B.: The Kannan's fixed point theorem in a cone rectangular metric space. J. Nonlinear Sci. Appl. **2**, 161–167 (2009)
150. Jleli M., Samet, B.: Remarks on G-metric spaces and fixed point theorems. Fixed Point Theory Appl. **2012**, 2012:210 (2011)
151. Kadeburg Z., Radenovič, S.: On generalized metric spaces: a survey. TWMS J. Pure Appl. Math. **5**, 3–13 (2014)
152. Kannan, R.: Some remarks on fixed points. Bull. Calcutta Math. Soc. **60**, 71–76 (1968)
153. Karapınar, E., Fulga, A., Alghamdi, M.: A common fixed-point theorem for iterative contraction of Seghal type. Symmetry **11**, 470 (2019). https://doi.org/10.3390/sym11040470
154. Karapınar, E., Shobkolaei, N., Sedghi, S., Vaezpour, S.M.: A common fixed point theorem for cyclic operators on partial metric spaces. Filomat. **26**(2), 407–414 (2012)
155. Karapınar, E., Taş K., Rakočević, V.: Advances on fixed point results on partial metric spaces. In: Tas, K., Tenreiro Machado, J.A., Baleanu, D. (eds.) Mathematical Methods in Engineering: Theory. Springer, pp. 1–59 (2018)
156. Karapınar, E.: A new non-unique fixed point theorem. J. Appl. Funct. Anal. **7**, 92–97 (2012)
157. Karapınar, E.: A note on common fixed point theorems in partial metric spaces. Miskolc Math. Notes. **12**, 185–191 (2011)
158. Karapınar, E., Du, W.S.: A note on b-cone metric and its related results: generalizations or equivalence? Fixed Point Theory Appl. **2013**, 2013:210 (2013)
159. Karapınar, E., Agarwal, R.P.: A note on Ćirić type non-unique fixed point theorems. Fixed Point Theory Appl. **2017** 2017:20 (2017)
160. Karapınar, E.: Discussion on (α, ψ) contractions on generalized metric spaces. Abstr. Appl. Anal. **2014**, Article ID 962784 (2014)
161. Karapınar, E., Piri, H., AlSulami, H.: Fixed points of generalized f-Suzuki type contraction in complete b-metric spaces. Discret. Dyn. Nat. Soc. **2015**, Article ID 969726 (2015)
162. Karapınar, E., Aydi, H., Samet, B.: Fixed points for generalized (alpha,psi)-contractions on generalized metric spaces. J. Inequal. Appl. **2014** 2014:229 (2014)

163. Karapınar, E., Erhan, I.M.: Cyclic contractions and fixed point theorems. Filomat. **26**(4), 777–782 (2012)
164. Karapınar, E., Erhan, I.M.: Fixed point theorems for operators on partial metric spaces. Appl. Math. Lett. **24**, 1900–1904 (2011)
165. Karapınar, E., Erhan, I.M.: Fixed point theorem for cyclic maps on partial metric spaces. Appl. Math. Inf. Sci. **6**, 239–244 (2012)
166. Karapınar, E.: Ćirić types non-unique fixed point theorems on partial metric spaces. J. Nonlinear Sci. Appl. **5**, 74–83 (2012)
167. Karapınar, E.: Ćirić types non-unique fixed point results: a review. Appl. Comput. Math. **1**, 3–21 (2019)
168. Karapınar, E.: Fixed points results for α-admissible mapping of integral type on generalized metric spaces. Abstr. Appl. Anal. **2014**, Article Id: 141409 (2014)
169. Karapınar, E.: Fixed points results via simulation functions. Filomat. **30**(8), 2343–2350 (2016)
170. Karapınar, E.: Fixed point theory for cyclic weak ϕ-contraction. Appl. Math. Lett. **24**(6), 822–825 (2011)
171. Karapınar, E., Sadaranagni, K.: Fixed point theory for cyclic (ϕ-ψ)-contractions. Fixed Point Theory Appl. **2011**, 2011:69 (2011)
172. Karapınar, E.: Fixed point theorems in cone Banach spaces. Fixed Point Theory Appl. **2009**, Article ID 609281 (2009). https://doi.org/10.1155/2009/609281
173. Karapınar, E., Erhan, I.M., Ozturk, A.: Fixed point theorems on quasi-partial metric spaces. Math. Comput. Model. **57**(9–10), 2442–2448 (2013)
174. Karapınar, E., Samet, B.: Generalized $\alpha - \psi$- contractive type mappings and related fixed point theorems with applications. Abstr. Appl. Anal. **2012**, Article ID 793486 (2012). https://doi.org/10.1155/2012/793486
175. Karapınar, E.: Generalizations of Caristi kirk's theorem on partial metric spaces. Fixed Point Theory Appl. **2011**, 2011:4 (2011)
176. Karapınar, E., Romaguera, S.: Nonunique fixed point theorems in partial metric spaces. Filomat. **27**(7), 1305–1314 (2013)
177. Karapınar, E.: Revisiting the Kannan type contractions via interpolation. Adv. Theory Nonlinear Anal. Appl. **2**(2), 85–87 (2018)
178. Karapınar, E., Yuksel, U.: Some common fixed point theorems in partial metric spaces. J. Appl. Math. **2011**, Article ID 263621 (2011)
179. Karapınar, E.: Some nonunique fixed point theorems of Ćirić type on cone metric spaces. Abstr. Appl. Anal. **2010**, Article ID 123094 (2010)
180. Karapınar, E.: Some fixed points results on Branciari metric spaces via implicit functions. Carpathian J. Math. **31**, 339–348 (2015)
181. Karapınar, E.: Some Fixed Point Theorems on the class of comparable partial metric spaces on comparable partial metric spaces. Appl. General Topol. **12**(2), 187–192 (2011)
182. Kutbi, M.A., Karapınar, E., Ahmed, J., Azam, A.: Some fixed point results for multi-valued mappings in b-metric spaces. J. Inequal. Appl. **2014**, 2014:126 (2014)
183. Karapınar, E., Alsulami, H.H., Noorwali, M.: Some extensions for Geragthy type contractive mappings. J. Inequal. Appl. **2015**, 2015:303 (2015)
184. Karapınar, E., Pitea, A.: On (α, ψ)-Geraghty contraction type mappings on quasi-Branciari metric spaces. J. Nonlinear Convex Anal. **17**, 1291–1301 (2016)
185. Kadeburg, Z., Radenoviè, S.: On generalized metric spaces: a survey. TWMS J. Pure Appl. Math. **5**, 3–13 (2014)
186. Karapınar, E.: On (α, ψ) contractions of integral type on generalized metric spaces. In: Mityushevand, V., Ruzhansky, M. (eds.) Proceedings of the 9th ISAAC Congress. Springer, Krakow, Poland (2013)

187. Karapınar, E., Kuman, P., Salimi, P.: On $\alpha - \psi$-Meri-Keeler contractive mappings. Fixed Point Theory Appl. **2013**, 2013:94 (2013)

188. Karapınar, E.: Weak ϕ-contraction on partial metric spaces. J. Comput. Anal. Appl. **14**, 206–210 (2012)

189. Karapınar, E.: Weak ϕ-contraction on partial metric spaces. J. Comput. Anal. Appl. **14**(2), 206–210 (2012)

190. Karapınar, E., Czerwik, S., Aydi, H.: (α, ψ)-Meir-Keeler contraction mappings in generalized b-metric spaces. J. Funct. Spaces. **2018**, Article ID 3264620 (2018)

191. Khan, M.S.: A fixed point theorem for metric spaces. Rend. Inst. Math. Univ. Trieste., Vol VIII, Fase. **10**, 1–4 (1976)

192. Khan, M.S.: Some fixed point theorems. Indian J. Pure Appl. Math. **8**, 1511–1514 (1977)

193. Khan, M.S., Swaleh, M., Sessa, S.: Fixed point theorems by altering distances between the points. Bull. Aust. Math. Soc. **30**(1), 1–9 (1984)

194. Khojasteh, F., Shukla, S., Radenović, S.: A new approach to the study of fixed point theorems via simulation functions. Filomat **29**(6), 1189–1194 (2015)

195. Kikina, L., Kikina, K.: A fixed point theorem in generalized metric space. Demonstratio Math. **XLVI**, 181–190 (2013)

196. Kincses, J., Totik, V.: Theorems an counterexamples on contractive type mappings. Math. Balkanica (N.S.) **4**(1), 69–90 (1990)

197. Kirk, W.A., Srinivasan, P.S., Veeramani, P.: Fixed points for mappings satisfying cyclical contractive conditions. Fixed Point Theory. **4**(1), 79–89 (2003)

198. Kramosil, O., Michalek, J.: Fuzzy metric and statistical metric spaces. Kybernetika **11**, 326–334 (1975)

199. Krasnosel'skiĭ, M.A., Zabrejko, P.P.: Geometric Methods of Nonlinear Analysis. Nauka, Moscow (1975) (in Russia)

200. Krasnoselskii, M.A.: Two remarks about the method of successive approximations. Uspehi Mat. Nuuk. **19**, 123–127 (1955)

201. Kutbi, M.A., Karapınar, E., Ahmed, J., Azam, A.: Some fixed point results for multi-valued mappings in b-metric spaces. J. Inequal. Appl. **2014**, 2014:126 (2014)

202. Künzi, H.P.A., Pajoohesh, H., Schellekens, M.P.: Partial quasi-metrics. Theoret. Comput. Sci. **365**, 237–246 (2006)

203. Kolmogorov, A.N., Fomin, S.V.: Elements of the Theory of Functions and Functional Analysis, vol. I. Graylock Press, Rochester, New York, Metric and Normed Spaces (1957)

204. Kopperman, R., Matthews, S.G., Pajoohesh, H.: What do partial metrics represent?, Spatial representation: discrete vs. continuous computational models, Dagstuhl Seminar Proceedings, No. 04351, Internationales Begegnungs- und Forschungszentrum für Informatik (IBFI), Schloss Dagstuhl, Germany (2005)

205. Liu, Z.Q.: On Ćirić type mappings with a nonunique coincidence points. Mathematica (Cluj) **35**(58), 221–225 (1993)

206. Liu, Z., Guo, Z., Kang, S.M., Lee, S.K.: On Ćirić type mappings with nonunique fixed and periodic points. Int. J. Pure Appl. Math. **26**(3), 399–408 (2006)

207. Liouville, J.: Second mémoire sur le développement des fonctions ou parties de fonctions en séries dont divers termes sont assujettis á satisfaire a une m eme équation différentielle du second ordre contenant un paramétre variable. J. Math. Pure et Appi. **2**, 16–35 (1837)

208. Maia, M.G.: Un'osservazione sulle contrazioni metriche. (Italian) Rend. Sem. Mat. Univ. Padova **40**, 139–143 (1968)

209. Matkowski, J.: Integrable solutions of functional equations. Diss. Math. **127**, Warsaw (1975)

210. Matthews, S.G.: Partial metric topology, Research Report 212. University of Warwick, Department of Computer Science (1992)

211. Matthews, S.G.: Partial metric topology, Proc. 8th Summer Conference on General Topology and Applications, Ann. New York Acad. Sci. **728**, 183–197 (1994)
212. Matkowski, J.: Fixed point theorems for mappings with a contractive iterate at a point. Proc. Am. Math. Soc. **62**(2) (1977)
213. Meir, A., Keeler, E.: A theorem on contraction mappings. J. Math. Anal. Appl. **28**, 326–329 (1969)
214. Menger, K.: Statistical metrics. Proc. Nat. Acad. Sci. USA **28**, 535–537 (1942)
215. Meyers, P.R.: A converse to Banach's contraction theorem. J. Res. Nat. Bur. Standards Sect. B **71B**, 73–76 (1967)
216. Mihet, D.: On Kannan fixed point principle in generalized metric spaces. J. Nonlinear Sci. Appl. **2**, 92–96 (2009)
217. Mitrović, Z.D.: A fixed point theorem for mappings with a contractive iterate in rectangular b-metric spaces. Matematiqki Vesnik **70**(3), 204–210 (2018)
218. Mizoguchi, N., Takahashi, W.: Fixed point theorems for multivalued mappings on complete metric spaces. J. Math. Anal. Appl. **141**, 177–188 (1989)
219. Mustafa, Z., Sims, B.: A new approach to generalized metric spaces. J. Nonlinear Convex Anal. **7**(2), 289–297 (2006)
220. Nemytskii, V.V.: The fixed point method in analysis. Usp. Mat. Nauk **1**, 141–174 (1936). (in Russian)
221. Nieto, J.J., Rodríguez-López, R.: Contractive mapping theorems in partially ordered sets and applications to ordinary differential equations. Order **22**, 223–239 (2005)
222. Oltra, S., Valero, O.: Banach's fixed point theorem for partial metric spaces. Rend. Istid. Math. Univ. Trieste **36**, 17–26 (2004)
223. Pachpatte, B.G.: On Ćirić type maps with a nonunique fixed point. Indian J. Pure Appl. Math. **10**, 1039–1043 (1979)
224. Pacurar, M., Rus, I.A.: Fixed point theory for cyclic φ-contractions. Nonlinear Anal. **72**, 1181–1187 (2010)
225. Paesano, D., Vetro, P.: Suzuki's type characterizations of completeness for partial metric spaces and fixed points for partially ordered metric spaces. Topol. Appl. **159**, 911–920 (2012)
226. Pant, R., Lohani, A., Jha, K.: A history of fixed point theorems. Ganita-Bharati Bull. Soc. Hist. Math. Ind. **24**, no. 1–4, 147–159 (2002)
227. Petric, M.A.: Some results concerning cyclical contractive mappings. General Math. **18**(4), 213–226 (2010)
228. Picard, E.: Memoire sur la theorie des equations aux derivees partielles et la methode des approximations successives. J. Math. Pures et Appl. **6**, 145–210 (1890)
229. Poincaré, H.: Surless courbes define barles equations differentiate less. J.de Math. **2**, 54–65 (1886)
230. Popescu, O.: Some new fixed point theorems for α–Geraghty-contraction type maps in metric spaces. Fixed Point Theory Appl. **2014**, 190 (2014)
231. Proinov, P.D.: A generalization of the Banach contraction principle with high order of convergence of successive approximations. Nonlinear Anal. (TMA) **67**, 2361–2369 (2007)
232. Proinov, P.D.: New general convergence theory for iterative processes and its applications to Newton Kantorovich type theorems. J. Complex. **26**, 3–42 (2010)
233. Rakotch, E.: A note on contractive mappings. Proc. Am. Math. Soc. **13**, 459–465 (1962)
234. Rad, G.S., Shukla, S., Rahimi, H.: Some relations between n-tuple fixed point and fixed point results. RACSAM **109**, 471–481 (2015)
235. Ran, A.C.M., Reurings, M.C.B.: A fixed point theorem in partially ordered sets and some applications to matrix equations. Proc. Am. Math. Soc. **132**, 1435–1443 (2003)
236. Ray, B.K., Rhoades, B.E.: Fixed point theorems for mappings with a contractive iterate. Pac. J. Math. **71** (1977)

237. Reich, S.: Fixed point of contractive functions. Boll. Un. Mat. Ital. **4**(5), 26–42 (1972)
238. Reich, S.: Kannan's fixed point theorem. Boll. Un. Mat. Ital. **4**(4), 1–11 (1971)
239. Reich, S.: Some remarks concerning contraction mappings. Can. Math. Bull. **14**, 121–124 (1971)
240. Rhoades, B.E.: A comparison of various definitions of contractive mappings. Trans. Am. Math. Soc. **224**, 257–290 (1977)
241. Rhoades, B.E.: Some theorems on weakly contractive maps. Nonlinear Anal. (TMA) **47**, 2683–2693 (2001)
242. Roldán-López-de-Hierro, A.F., Karapınar, E., Roldán-López-de-Hierro, C., Martínez-Moreno, J.: Coincidence point theorems on metric spaces via simulation functions. J. Comput. Appl. Math. **275**, 345–355 (2015)
243. Roldán, A., Martinez-Moreno, J., Roldán, C., Karapınar, E.: Multidimensional fixed point theorems in partially ordered complete partial metric spaces under (psi,varphi)-contractivity conditions. Abstr. Appl. Anal. **2013** Article Id: 634371 (2013)
244. Roldán, A., Martinez-Moreno, J., Roldán, C., Karapınar, E.: Some remarks on multidimensional fixed point theorems, Fixed Point Theory **15**(2), 545–558 (2014)
245. Romaguera, S.: A Kirk type characterization of completeness for partial metric spaces. Fixed Point Theory Appl. **2010**, Article ID 493298 (2010)
246. Romaguera, S., Schellekens, M.: Duality and quasi-normability for complexity spaces. Appl. General Topology. **3**, 91–112 (2002)
247. Romaguera, S., Schellekens, M.: Partial metric monoids and semivaluation spaces. Topol. Appl. **153**, 948–962 (2005)
248. Romaguera, S., Valero, O.: A quantitative computational model for complete partial metric spaces via formal balls. Math. Struct. Comput. Sci. **19**, 541–563 (2009)
249. Romaguera, S.: Fixed point theorems for generalized contractions on partial metric spaces. Topol. Appl. **159**, 194–199 (2012)
250. Romaguera, S.: Matkowski's type theorems for generalized contractions on (ordered) partial metric spaces. Appl. General Topol. **12**, 213–220 (2011)
251. Rus, I.A.: Principles and Applications of the Fixed Point Theory (in Romanian). Editura Dacia, Cluj-Napoca (1979)
252. Rus, I.A.: Generalized Contractions and Applications. Cluj University Press, Cluj-Napoca, Romania (2001)
253. Rus, I.A.: Cyclic representations and fixed points. Ann. T. Popoviciu, Seminar Funct. Eq. Approx. Convexity **3**, 171–178 (2005)
254. Samet, B., Vetro, C., Vetro, P.: Fixed point theorem for $\alpha - \psi$ contractive type mappings. Nonlinear Anal. **75**, 2154–2165 (2012)
255. Samet, B., Vetro, C., Vetro, F.: From metric spaces to partial metric spaces. Fixed Point Theory Appl. **2013**(5), 1–11 (2013)
256. Samet, B., Vetro, C., Vetro, F.: Remarks on G-metric spaces. Int. J. Anal. **2013**, Article ID 917158 (2013)
257. Samet, B.: Discussion on: a fixed point theorem of Banach-Caccioppoli type on a class of generalized metric spaces by A. Branciari. Publ. Math. Debrecen. **76**(4), 493–494 (2010)
258. Samet, B., Rajović, M., Lazović R., Stoiljković, R.: Common fixed point results for nonlinear contractions in ordered partial metric spaces. Fixed Point Theory Appl. **2011**, (71) (2011)
259. Samet, B., Karapınar, E., Aydi, H., Rajic, V.: Discussion on some coupled fixed point theorems. Fixed Point Theory Appl. **2013**, 50 (2013)
260. Schauder, J.: Der Fixpunktsatz in Funktionalraumen. Studia Math. **2**, 171–180 (1930)
261. Schellekens, M.P.: A characterization of partial metrizability: domains are quantifiable. Theoret. Comput. Sci. **305**, 409–432 (2003)

262. Schellekens, M.P.: The correspondence between partial metrics and semivaluations. Theoret. Comput. Sci. **315**(1), 135–149 (2004)
263. Sehgal, V.M.: Some fixed and common fixed point theorems in metric spaces. Can. Math. Bull. **17**(2), 257–259 (1974)
264. Sehgal, V.M.: A fixed point theorem for mappings with a contractive iterate. Proc. Am. Math. Soc. **23**, 631–634 (1969)
265. Shatanawi, W., Samet, B., Abbas, M.: Coupled fixed point theorems for mixed monotone mappings in ordered partial metric spaces. Math. Comput. Model. **55**(3–4), 680–687 (2012)
266. Shobkolaei, N., Vaezpour, S.M., Sedghi, S.: A common fixed point theorem on ordered partial metric spaces. J. Basic. Appl. Sci. Res. **1**(12), 3433–3439 (2011)
267. Singh, K.L.: Fixed-point theorems for contractive-type mappings. J. Math. Anal. Appl. **72**, 283–290 (1979)
268. Suzuki, T.: A generalized Banach contraction principle that characterizes metric completeness. Proc. Am. Math. Soc. **163**, 1861–1869 (2008)
269. Suzuki, T.: A new type of fixed point theorem on metric spaces. Nonlinear Anal. **71**, 5313–5317 (2009)
270. Suzuki, T.: Fixed point theorems and convergence theorems for some generalized nonexpansive mappings. J. Math. Anal. Appl. **340**, 1088–1095 (2008)
271. Tarski, A.: A lattice theoretical fixpoint theorem and its applications. Pac. J. Math. **5**, 285–309 (1955)
272. Valero, O.: On Banach fixed point theorems for partial metric spaces. Appl. General Topol. **6**(2), 229–240 (2005)
273. Vetro, C., Vetro, F.: Metric or partial metric spaces endowed with a finite number of graphs: a tool to obtain fixed point results. Topol. Appl. **164**, 125–137 (2014)
274. Vetro, C., Vetro, F.: Common fixed points of mappings satisfying implicit relations in partial metric spaces. J. Nonlinear Sci. Appl. **6**(3), 152–161 (2013)
275. Vetro, F., Radenović, S.: Nonlinear ψ-quasi-contractions of Ćirić-type in partial metric spaces. Appl. Math. Comput. **219**(4), 1594–1600 (2012)
276. Valero, O.: On Banach fixed point theorems for partial metric spaces. Appl. Gen. Topol. **6**, 229–240 (2005)
277. Waszkiewicz, P.: Partial metrisability of continuous posets. Math. Struct. Comput. Sci. **16**(2), 359–372 (2006)
278. Waszkiewicz, P.: Quantitative continuous domains. Appl. Categ. Struct. **11**, 41–67 (2003)
279. W Wilson, W.A.: On semimetric spaces. Am. J. Math. **53**(2), 361–373 (1931)
280. Zamfirescu, T.: Fixed point theorems in metric spaces. Arch. Math. (Basel) **23**, 292–298 (1972)
281. Zhang, F., Kang, S.M., Xie, L.: Ćirić type mappings with a non-unique coincidence points. Fixed Point Theory Appl. **6**, 187–190 (2007)

Index

E. Karapınar and R. P. Agarwal, *Fixed Point Theory in Generalized Metric Spaces*,
Synthesis Lectures on Mathematics & Statistics,
https://doi.org/10.1007/978-3-031-14969-6

Printed in the United States
by Baker & Taylor Publisher Services